Corrales Bosque
Corrales NM

Sept 29, 1996

Michael,
Keep up the superb work!
It's appreciated.
Max

Caring
for
Creation

Yale
University
Press

New
Haven
and
London

Caring for Creation

Max Oelschlaeger

An Ecumenical Approach to the Environmental Crisis

Published with assistance from the
Mary Cady Tew Memorial Fund.

Designed by Deborah Dutton.
Set in Joanna type by Keystone Typesetting, Inc.
Printed in the United States of America by Vail-Ballou Press, Binghamton, New York.

Oelschlaeger, Max.
Caring for creation : an ecumenical approach to the environmental crisis / Max Oelschlaeger.
p. cm.
Includes bibliographical references and index.
ISBN 0-300-05817-9 (cloth: alk. paper)
0-300-06645-7 (pbk.: alk. paper)
1. Human ecology—Religious aspects—Christianity. 2. Human ecology—Religious aspects—Judaism. I. Title.
BT695.5.034 1994
261.8′362—dc20 93-23215
CIP

A catalogue record for this book is available from the British Library.

The paper in this book meets the guidelines for permanence and durability of the Committee on Production Guidelines for Book Longevity of the Council on Library Resources.

10 9 8 7 6 5 4 3 2

for Mary Lee and Peter Max

Contents

Acknowledgments

I am indebted to those individuals whose texts inspired me and to the readers who helped me develop the argument. The former will be evident from the book itself. The latter include scientists and religious believers, as well as philosophers and political scientists. I am especially obliged to my colleagues Pete Gunter, Joe Barnhart, Susan Bratton, and Gene Hargrove; collectively they helped me refine philosophical and theological claims, leading to a more forcefully stated account of the relation between environmental ethics and religious faith. Gunter, an environmentalist for nearly four decades, critiqued every page of the manuscript, some on more than one occasion. Barnhart, a philosopher of religion, provided many interesting suggestions, some of which I was unable to incorporate. Bratton, now a colleague of mine, read the manuscript from the vantage point of an ecologist who also works at the intersection of religion and environmental ethics. Hargrove, the editor of *Environmental Ethics*, first gave me the idea of writing such a book. I also thank the ecologists Tom Waller and Ken Stewart, and the deep ecologist Dolores LaChapelle, who brought to their readings different faith commitments and sharpened my description of the relation between ecological and religious discourse. Bob Paehlke helped improve the political side of the argument, and Jay McDaniel offered thoughtful comments on ecotheology. Mark Dorfman and Mary Oelschlaeger offered useful suggestions for making the text more readable. Two anonymous referees contributed valuable criticism that I incorporated in the final writing of the manuscript. Finally, I am again indebted to the executive editor of Yale University Press, Charles Grench, to his associate Otto Bohlmann, and to Laura Jones Dooley for her gentle yet incisive and imaginative copy editing.

Introduction

"Where there is no vision," the Bible tells us, "the people perish" (Prov. 29:18, KJV). The question of our age is whether such vision will emerge in time to stem the tides of ecological destruction, social injustice, and war.—Jay B. McDaniel, *Of God and Pelicans: A Theology of Reverence for Life*

In a sense, this book is a confession, though not exactly in the tradition of Augustine and other Christians. For most of my adult life I believed, as many environmentalists do, that religion was the primary cause of ecological crisis. I also assumed that various experts had solutions to environmental malaise. I was a true believer: If only people would listen to the ecologists, economists, and others who made claims that they could "manage planet Earth," we would all be saved. I lost that faith by bits and pieces, especially through the demystification of two ecological problems—climate heating and extinction of species—and by discovering the roots of my prejudice against religion. That bias grew out of my reading of Lynn White's famous essay blaming Judeo-Christianity for the environmental crisis. In some ways this book can be read as accounting for my change of mind or "conversion experience." The story begins in chapter 1 with the debate over Lynn White and ends five chapters later with a theory for interpreting religion in a time of ecocrisis. But the facts, or at least some basic information about climate heating and the extinction of species, are the place to begin.

Though the scientific community cannot exactly model the climatological future, and even though climatologists may disagree, there is no mystery today about climate heating, the so-called global greenhouse.[1] The primary physical causes are (1) the carbon dioxide that comes from automotive emissions and the use of hydrocarbons to fuel energy-intensive industrialized economies and (2) deforestation, especially the destruction of rainforests. These physical variables in turn reflect economic and other sociocultural variables, such as the corporations that produce energy products and related services and the consumers who use them. Yet the energy leviathans of the modern age—Texaco and Exxon—are no more evil, malevolent corporations embarked on a mission to destroy the earth than are the individuals who drive automobiles to the office or the mall. We all, corporately and individually, are simply living within the existing social matrix—the market economy, or utopian capitalism, as it is more accurately called.

Similarly, there is no enigma about the loss of biodiversity.[2] The relentless growth of human population and the economic development associated with it have either claimed the habitats necessary for wild species to survive or polluted those habitats in life-endangering ways. In technical terms, the causes of the twentieth-century extinction of species are anthropogenic—of human origin. As with climate heating, modern technology and the dynamics of the market drive the process. The outcome is the relentless conversion of habitat and appropriation of water and other resources for human purposes—those of both producers and consumers. There is little promise of either a strictly political or an economic solution to the problem, since politicians and markets respond to the short-term imperatives of human society rather than the long-term cycles of nature. Nowhere does the preservation of either ecosystems or species fit into the dominant cultural calculus of preference.

And so on through the permutations of possible ecocatastrophe.[3] We are the environmental crisis, and it is primarily our philosophies, economies, and governments that motivate and direct the devastating onslaught against the earth. However, the pervasive idea that there are "green saviors" who occupy the environmental high ground and "evil exploiters" who are rapaciously abusing nature is not useful. For one reason, human beings are powerfully motivated, indeed, genetically impelled, to survive in the short-run whatever the long-term costs. Ecological problems created by present ways of living will not be solved until meaningful alternatives, such as actual substitutes for fossil fuels, like solar energy, are institutionalized. For another reason, creating a binary opposition between "good guys" who protect nature and "bad guys" who destroy it assigns blame to some groups and excuses others. This dichotomization does not go very far in helping Americans achieve the solidarity needed to change behavior through political action and economic restructuring. Chrysler and General Motors and you and I are caught up together in modern society, acting out our roles in a cultural script we did not write.

I do not know a single environmentalist who lives outside the prevailing social matrix. Consider transportation: all of my green friends, many of whom are an inspiration to me, have cars (some have more than one). They all fly on commercial airlines to conferences and workshops around the world. Many commute long distances to their jobs in the academy and elsewhere. Given such behavior, how can they claim to know better? Gary Snyder, a leading proponent of deep ecology and Pulitzer prize-winning nature poet, was confronted with this conundrum by a questioner at a workshop I attended in 1989. "How can you justify jetting around from Alaska to Colorado to California?" this person wondered. "Don't you see that the consequences of your actions, however small, are just like everyone else's? and that in the aggregate they make up ecocrisis?" Snyder gave, I thought, an honest answer. "We're all part of the problem," he said. "I'm not the answer. Or the final destination. Maybe I'm a signpost. For the earth to survive we'll have to find the route together." Aldo Leopold, founder of the Wilderness Society and perhaps the most influential environmental ethicist in American history, would agree. Leopold ([1949] 1970) confessed that he, and everyone else who becomes environmentally aware, must realize that we live in a world of wounds. We all contribute to the process of injury.

The point of my confession, then, is that environmentalists like myself can be sanctimonious. People who portray themselves as nature's champions and corporations as evil villains are not always contributing positively to efforts that lead the way beyond ecocrisis. Writing and teaching about green issues are in one sense part of the solution, but most ecological writers and teachers live in ways that, whatever the green gloss on the surface, are environmentally flawed. My life is typically paradoxical, since on the one hand I am an environmentalist, teach in an internationally recognized environmental philosophy program, and have written and edited books on ecological issues. I live in a passive solar energy house, eat low on the food chain, save virtually anything that can be recycled, and keep my thermostat at energy-conserving levels. But I own—through a university retirement system—a considerable portfolio of stocks heavily invested in Fortune 500 companies; the size of my pension will depend on the continued growth and profitability of these corporations. I am also a typical consumer of energy and related products. My car's engine, its air conditioner, and even the food I eat undermine global ecology. Like all members of the social matrix, I am more part of the problem than the solution. My high-energy, high-consumption way of life undercuts the biophysical processes that sustain the Creation, including civilization itself.

The Paradox of Environmentalism

Ironically, the global ecocrisis continues to worsen despite people's efforts to respond—a situation I call the *paradox of environmentalism.* World population steadily grows, global levels of atmospheric CO_2 continue to rise, and thousands of species are driven into extinction each decade. An account of environmental degradation could fill this book. Americans have tried to respond to ecocrisis, largely through initiatives that have created bureaucracies and communities of experts to administer environmental laws and policies. These efforts are so pervasive that the "green collar" work force is now recognized alongside the traditional designations of blue and white collar. And yet, as I have confessed, I have no faith that experts can *in the present context* meaningfully contribute to solutions. In truth, the experts, too, are more part of the problem than the solution.

Environmental or green politics offers, at this juncture, little encouragement, since any celebration of victory, as in the continued protection of the Arctic National Wildlife Refuge (1991), is immediately dampened by the defeat of such policy initiatives as the so-called Big Green referendum in California (1990). Environmentalists appear unable to create a collective vision of a green or sustainable society; a piecemeal, ad hoc approach—a "put out the fire" mentality—governs. To some environmentalists the government itself is the problem; voters have acquired a new skepticism about self-proclaimed environmental presidents. Distrust would be even more widespread if the public knew that the fox has been sent to guard the henhouse: the federal government is the nation's number one polluter and destroyer of wildlife habitat. Federal agencies and facilities may violate environmental statutes with impunity, and states have no recourse to prosecute or sue for noncompliance. Many agencies charged with administering the law are underfunded. The entire Environmental Protection Agency (EPA) budget for implementing the Pollution Prevention Act (1990), for example, is a mere sixteen million dollars per year; this is especially ironic since the bill is predicated on the cost-effective nature of voluntary pollution prevention compared with the cost-ineffective nature of mandatory clean-up (for example, the EPA's Superfund).

The deeper issue is that the notion of expert solutions is symptomatic of the disease, not part of the cure. American society worked originally because ordinary people, rooted in local culture, participated in the important decisions of governance. This is no longer so. Ordinary people, wherever they live, are removed from the decision-making processes that weave the basic socioeconomic fabric of their lives; elected officials and bureaucratic experts increasingly make decisions for them. Which is precisely the problem. These officials and experts *necessarily* serve the industrial growth system that creates and employs them even as that system undercuts the ecosystemic bases of sustainable civilization. Politicians and experts are incapable of dealing with the systemic causes of ecocrisis because they are too much a part of the system and machinery of the Great Society.

But all is not lost. Politicians and experts are not intrinsically evil. And the means to affect movement toward sustainability exists. I think of religion, or more specifically the church—both the public church and congregations of people or fellowships of believers gathered in places of worship, engaging in discourse about their responsibilities to care for creation within the context of their traditions of faith—as being more important in the effort to conserve life on earth than all the politicians and experts put together. The church may be, in fact, our last, best chance. My conjecture is this: *There are no solutions for the systemic causes of ecocrisis, at least in democratic societies, apart from religious narrative.*

Environmentalists generally, I think, will be skeptical of this claim. To them, religion is the cause or part of the cause of ecocrisis. "After all," they argue, "Judeo-Christians believe that they have dominion over the earth and do not believe that they are an integral part of biotic communities. And in any case, we need science, and especially new technologies, to solve environmental problems. How could religion be relevant?" I ask that readers who are dubious to give my thesis a fair chance—a hearing. My claim is not that religion alone can resolve the environmental crisis but that it has an irreplaceable function in the larger process. One role is helping people begin to change daily practices that have adverse ecological effects. A second purpose is to change the lay of the political landscape, primarily through the election of leaders who are genuinely rather than rhetorically responsive to ecocrisis.

Working through the political process, Americans could begin to change their own behavior and that of politicians to reflect our increasing grasp of ecological problems. Whatever the present situation, citizens can influence politics by shaping, especially through the ballot box, governmental policy (local, regional, state, and national). Consider one example: energy efficiency. By increasing energy efficiency through building construction codes, alternative energy technologies, and improved manufacturing and transportation techniques, the United States could reduce its CO_2 emissions by almost 50 percent while simultaneously improving its ability to sell goods internationally. Contrary to popular opinion and the scare tactics of the anti-environmental movement, ecologically informed policy-making makes good economic sense (unless economic theory is restricted conceptually to monetary measures of marketed consumption). Yet we do not elect leaders who promote energy efficiency. Clearly, the democratic consensus necessary to implement appropriate public policies does not now exist.

No one policy change will resolve ecocrisis. If energy efficiency leads to lower prices, then increased consumption might result, leading to the conversion of more resources and habitat into more goods to fuel the continued growth of population and the economy. If alternative energy technologies and more efficient use of present technologies are to help, additional measures that address population control, biodiversity, pollution abatement, and resource consumption are required. Still, as John Firor, former director of the Advanced Study Program at the National Center for Atmospheric Research argues, if we collectively try to *solve any one aspect of ecocrisis*, we will likely *solve all* of them, since they are intrinsically linked (1990, 109). In discussing problems of acid rain, climate heating, and atmospheric ozone loss, Firor points out that although solutions for each problem are characteristically considered alone, in truth they are "strongly interrelated, and all three are tightly connected to such national and global issues as the energy supply, Third World development, foreign trade, North-South equity, and the population explosion" (x).

If, in spite of two decades of widespread public concern, solidarity on environmental policies and leaders is more an ideal than an actuality, and if the experts give no indication of being able to redirect society toward sustainability, what is the alternative? Can it be that religion is essential to a green future? Can it be that religion might politically empower our leaders to implement policies that are scientifically informed and technologically appropriate? In my opinion the answer to these questions is yes, however strange that notion may seem to the skeptic. The reasons why I think that religious discourse might move us democratically toward sustainability are the substance of this book. Readers can judge for themselves whether the conjecture is cogent. Here let me say only that religion, and therefore its role in ecocrisis, has been inadequately understood by both its foes and most of its proponents.

Many readers will come to this book inclined to believe that I am either soft on or hostile toward religion. Let me affirm that my thesis has nothing to do with being either a believer or an unbeliever, or with theism and naturalism per se. Further, my argument is neutral between, for example, conservative Baptists and religious liberals. In context, it makes no difference whether readers believe that the Bible is the inerrant word of God or simply literature. If anything, the text is directed at both people of faith and those who question such belief, since they share a common environment. Of course, different faiths offer different resources for defining an environmental ethic. Some faith traditions have taken the lead over others in responding to ecocrisis and in developing an ecoethic. Yet in contemporary context the so-called greening of religion is unmistakably occurring.

In responding to papers I have presented on religion and ecology, and in reading drafts of these chapters, some environmentalists have accused me of ignoring Judeo-Christianity's culpability for the environmental crisis. Others have doubted that religious discourse has anything to do with finding the way out of ecocrisis, favoring more ecologically oriented education, more scientific data and analysis, and ultimately more ecotechnologies as the key.[4] Some religious conservatives, in contrast, have accused me of siding with godless scientists who provide the technology that exploits God's creation. The solution to ecocrisis, as they see it, is to accept the reality of the Fall—our sinful human nature. Therein, they claim, lies the possibility of transformation.

Through dialogue with my critics, both environmentalists and religious believers of many persuasions, a change has occurred. We have reached a consensus that, whatever religion's responsibility for ecocrisis has been or even might be, it is more important to emphasize the role religion can now play in helping society ameliorate that crisis. No doubt, disagreements remain; there is a diversity of opinion on some issues. As noted, some ecologists believe that I have not emphasized strongly enough the importance of educating lay publics about the environment. (On this subject, see chapter 5.) Some conservative Christians, on the other hand, believe that nothing, including ecological education, will make a difference unless we let Jesus into our lives. Clearly, this book will not satisfy everyone, and I have not tried to do so.

I have made every effort to avoid marginalizing any group or position. When it comes to protecting the future of life on this planet, *solidarity is more important than ideological supremacy*. I hope that *religious readers* will interpret this text as opening rather than closing a conversation, as an invitation to discuss the crucial public issue of the 1990s. I hope that *secular readers*, especially those from the natural sciences, will come to comprehend religious discourse in a different way—particularly as contributing to the creation of a cultural context in which, for example, environmental education and ecotechnologies can make a difference. I might also point out that many within the community of natural scientists have strong religious commitments. And there is at least one international organization, the Joint Appeal by Science and Religion for the Environment, that brings together scientists and religious believers in conversation about ecocrisis. More than thirteen hundred representatives attended its meeting in Moscow in January 1990 (addressed in a plenary session by Mikhail Gorbachev).

"Readers with attitudes" also include some of my academic colleagues, ecophilosophers, otherwise known as environmental ethicists. Many of them regard as irrational the idea that religion is essential to resolving the ecocrisis. Such a belief is an occupational hazard of *systematic philosophers*, who think of themselves as the ultimate experts—the providers of master theories (totalizing discourses) that provide a decision-making matrix for the rational solution of environmental problems. In the main, environmental ethics is a technical field for intellectuals, bristling with competing, esoteric theories and arcane, scholarly terminology that only experts and their acolytes can decipher. Fortunately, in spite of arguments for moral monism and master narratives, a new strain of pragmatism is beginning to transform environmental ethics, fostering a healthy skepticism of the claims of systematic ecophilosophers.

Overview

Caring for Creation examines a variety of religious traditions, including Goddess feminism and native American traditions, in making the case. The primary emphasis, however, is on Judeo-Christianity, since the United States was a biblical culture in its inception and—broadly construed—remains so today.

This does not mean that for practical purposes Judeo-Christian stories exclude non–Judeo-Christian stories. Further, within Judeo-Christianity no single faith is favored over any other. A wide spectrum of possibilities from conservative to liberal and even radical is examined.

Obviously, people of faith—in the broadest possible sense, both clergy and lay people—who believe that their religion relates to life itself constitute a vital part of my audience. Any claim that religious faith has nothing or little to do with life is self-defeating and therefore not a serious argument against my thesis. Such a claim also ignores solid evidence that religion remains a potent factor in the lives of most Americans. My argument encourages people of faith both within and outside the Judeo-Christian mainstream to reconsider the role of their religion in a time of ecocrisis. Even though there are grounds for disagreements among us, the concern I have is for the common good.

My argument may be novel, but it does not break new methodological or theological ground per se. I follow the lead of those who argue that the Bible is the Great Code apart from which the existence of Western culture becomes almost incomprehensible—as implied in the oxymoron "plotless story." Although I use a sociolinguistic method, my primary concern is not sociological, linguistic, or philosophical theory but social practice. Like many others, I believe that some aspects of our Enlightenment philosophy are less useful than was once thought. Primarily, this is because that master narrative expresses itself through the language of utilitarian individualism—a language that hinders us in developing any sense of either community or social goals (citizen preferences) outside the market. My argument draws on an array of materials concerned with religion specifically and the environmental crisis generally, ranging freely across disciplines as diverse as the sociology of religion, conservation biology, biblical hermeneutics, systems ecology, theology, critical rhetoric, climatology, and political science.

It is important to make clear a few presuppositions at the outset; later sections of the text will provide sustaining arguments. I assume that solutions to ecocrisis entail the democratic articulation of social preferences that cannot be expressed through the market. Solidarity on these preferences is essential. I also assume that human beings, living out their lives within a social matrix that creates ecocrisis, are more usefully described as storytelling culture-dwellers than as rational agents seeking ultimate knowledge of timeless foundations. That story—that thinking animals can have ultimate knowledge—begins with the philosophical narratives of the Presocratics and the Greeks, especially Plato and Aristotle. By assuming that human beings are storytelling culture-dwellers

I abandon the culturally dominant understanding of knowledge as representation and embrace the notion that in a time of ecological crisis solidarity is more important than "objectivity"—that is, objectivity understood as correspondence between the ideas of thinking subjects and an external, objective, timeless reality. *Caring for Creation* should be construed as recontextualization, as cutting us free from the dominance of legitimating narratives that reflect the modern paradigm and opening up the possibility of a genuine conversation about the possible forms of social existence. Chapter 3 argues that such an approach is consistent with the need to bring scientific information to bear on ecological problem solving. Chapter 6 contends that such an approach is consistent with—indeed, grows out of—the biblical tradition.

My assumption that human beings are usefully described as storytelling culture-dwellers should not be interpreted as closing off alternative claims or descriptions, such as those made by metaphysical realists, including both scientific and religious realists, that human beings can have ultimate knowledge. Metaphysical realism, either the scientific version, with its belief in an external, objective reality entirely independent of human beings that is known through science, or the religious version, with its belief in a God-who-is-there, has a long and venerable tradition. Further, it offers the comfort of certain conviction. My approach does not offer certain conviction; neither does it need to, for my aim is to offer an interpretation of ecocrisis through which competing claims to ultimate knowledge can be recognized without according a privileged position to any single claim. I make no attempt to adjudicate disputes between, for example, scientific and religious realists (discussed in chapter 2), nor between the competing claims of people of faith to ultimate knowledge. Scientists and religionists with a vested interest in defending their claims to ultimate knowledge see to that purpose themselves. My interest here is in the role of religion in helping Americans find solutions to ecocrisis. Practically considered, such solutions require politically coordinated, scientifically informed, and institutionally empowered actions that are consistent with our basic democratic freedoms and diversity of ultimate commitments. I do not make the claim that my perspective is a truer account of *the way things are.* I claim only that it is useful for the task at hand.

My approach suspends the instrumental view of language—that is, the belief that we are the masters and possessors of language—in favor of the idea that we belong to language. The book itself is the only defense I offer for such a position (though readers who wish to pursue the case might refer to the work of Hans-Georg Gadamer, Richard Rorty, and Paul Ricoeur among many others).

The point is that sociolinguistically framed, the Great Code, as Northrop Frye (1981) calls it, or our biblical tradition, as Robert Bellah and his coauthors (1985) term it, has more importance than most of us—living in our increasingly instrumental world and believing that we are the masters and possessors of language—comprehend. Another implication is that religious narrative is cognitively legitimate, and cannot be distinguished in principle from other story sources, such as science, economics, or philosophy. In short, no form of discourse is a priori privileged, although as a matter of sociological and political fact some forms of discourse are privileged over others. Asserting that no one form of discourse is privileged does not mean, for example, that religious discourse can replace ecological discourse. It does mean that God-talk, as religious narrative is sometimes called by sociolinguists, can not only frame but legitimately address such issues of ecological consequence as biodiversity, pollution, and population. As some environmentalists note, there is an inescapable moral dimension to ecocrisis: no strictly technological solutions exist. Thus, religious and scientific discourse might inform and reinforce each other in helping to transform culture.

A sociolinguistic perspective also entails the idea that, consistent with our republican tradition, individuals who enjoy communal benefits are obligated to participate in public discourse. In a time of crisis, such conversation may eventuate in solidarity, the widespread acknowledgment of an overriding public good. The republican tradition carries with it an obligation to act politically in order to realize citizen preferences. So conceptualized, religious discourse that concerns the theme of caring for creation is vital to our culture. It is, in fact, imperative if there is to be a *democratic transition* to sustainability, since religious discourse addresses the kinds of value considerations that are either neglected by or excluded from other forms of discourse. Contrary to the opinion of environmental philosophers, who believe that they alone can define an environmental ethic, a sociolinguistic analysis suggests that for most Americans an environmental ethic will either grow out of their religious faith or will not grow at all.

Religion also, and crucially, is the only form of discourse widely available to Americans (through the institution of the church) that expresses social interests going beyond the private interests articulated through economic discourse and institutionalized in the market. As coauthors Charles Lumsden and E. O. Wilson note (1981, 1983), and I use them deliberately as my example, since their argument comes not from a religious but a sociobiological perspective, organized religion is more concerned with the welfare of the group, the collective good, than any other institution. However, for religious discourse to address ecocrisis effectively, beyond its implications for redirecting personal behavior, it must articulate itself through the democratic process. Democracy today, by all available evidence, typically works better to serve narrowly defined private interests, particularly economic interests, than the common good. It sustains a "politics of interest" far more than a "politics of community" directed toward the public interest in sustainability and environmental preservation.[5] Alternatively stated, the democratic process is largely monolithic, dominated by the American Business Party, with its two wings, the Democrats and the Republicans. Utilitarian individualism is the lingua franca of society and of politics. Religious discourse, as I shall attempt to show, is perhaps the most promising way to expand our cultural conversation to include non-market values such as sustainability and to revitalize *citizen democracy*.

Environmentalism may become a potent political factor in the United States. Given the intractability of ecocrisis (for example, the global greenhouse will not spontaneously disappear), the issue of what kind of politics, and ultimately society, we are going to have is fundamental. Today, however, there is political indecisiveness—marked by a lack of authority that can come only from *moral purpose*. Americans seem to be caught between an unsustainable way of life and a future that is as yet powerless to be born—a postmodern age of sustainability. Public opinion polls indicate a widely shared perception that we have abused nature, that timely action is important, and that the politicians have done too little. But no effective political consensus exists. There is little mystery as to why the body politic has not moved more decisively. Americans fundamentally distrust institutions, which, in the traditional way of thinking, threaten individual freedom. Paradoxically, political and economic institutions are indispensable to effecting change. Concerned citizens are caught on the horns of a dilemma, confronting political choices that appear to be diametrically opposed. Environmental preservation clearly requires collective action through the political process. Yet any such initiative seems to constrain the market economy, itself a collective good reflecting the aggregate of private economic interests.

Part of our problem is that the Enlightenment definition of freedom as individual control over daily life and independence from governmental obstruction is increasingly inadequate. We need a more positive one. "Freedom," to quote the sociologist Robert Bellah and his colleagues, "must include the right to participate in the economic and political decisions that affect our lives. Indeed, the great classic criteria of a good society—peace, prosperity, freedom, justice—all depend today on a new experiment in democracy, a newly extended and enhanced set of democratic institutions, within which we citizens can better discern what we really want and what we ought to want to sustain a good life on this planet for ourselves and the generations to come" (1991, 9).

Religious discourse may help Americans come together as a nation and move toward sustainability because it converges, through the diversity of faith, on a common center of caring for creation. For religious discourse to be politically effective, both denominations and local churches—associations of congregations and pastors, lay people and clergy—must actively engage the religious implications of ecocrisis. Clearly, institutions are structured by and through language, through discourse that legitimates their creation, maintenance, and operation. Religious language, expressing itself in democratic context, can help create the political will to deal with ecocrisis, that is, it can help Americans change the institutionalized, self-perpetuating traditions that lead down an ecologically perilous path. Arguing that religious discourse is essential to creating a democratic consensus on an environmental agenda does not entail privileging any one claim to ultimate knowledge, that is, a so-called skyhook. Rather, every tradition of faith, consistent with its metaphysical claims (so-called transcendental signifieds that are beyond discussion), also offers "toeholds," especially new metaphors such as "caring for creation" that can lead to solidarity on an environmental agenda.[6] Such an agenda would possess the *moral authority* that reform environmentalism now lacks. As I attempt to confirm in chapter 4, every faith across the spectrum of religious belief can either find or has already found its way to an environmental ethic through a metaphor of caring for creation.[7]

Without religion, the environmental crisis will likely, and tragically, lead to *administrative despotism*. Commentators from across the sociopolitical spectrum, from deep ecologists on the political left to conservatives on the political right, fear the managed society. If such a society eventuates, it will testify to the inability to reach a consensus on caring for creation and thus the triumph of materialism—confirmation of a failure to approach ecocrisis in any way outside of utilitarian individualism. By default, the prevailing, *narrowly economic conception of the public good* will continue to dominate politics. Increasingly onerous, expensive, and bureaucratic forms of management will inevitably follow the environmentally destructive consequences of the pursuit of economic growth for the sake of growth alone (that is, the drive to maximize the Gross National Product without considering the qualitative mix of economic goods and services and their ecological costs as well as nonmonetary variables). The alternative is to reach through democratic consensus some conception of the public interest that guides the market toward sustainable processes of production and consumption. In ecological terms, the abuse of the earth is bad economics. By helping society move toward a sustainable economy, religion may lead to even more fundamental transitions in American culture. The question, then, is whether the American experiment will settle ever more firmly into administrative despotism or evolve into a democratic, sustainable society.

Several questions hover in the background of my argument. One is whether a genuinely ecumenical treatment of religion and environmentalism is possible. Most texts, even those proclaiming themselves ecumenical, are partisan: one perspective (story, narrative, theology) is privileged over another. The term *ecumenical*, however, does not imply total and complete agreement on every issue. Rather, ecumenical discourse does not marginalize any position on the basis of either favored presuppositions or final vocabularies used to judge others. My approach to religious discourse is one way to avoid partisanship, since the sociolinguist is reflexively embedded in the reality of language and thereby—to use David Tracy's (1987) words—ambiguity and plurality. My inquiry attempts in good faith not to marginalize any faith tradition. Rather than presupposing any one master narrative, I draw on a variety of religious stories in arguing that distinctive religions converge on the theme of caring for creation.

A second question involves assessing the degree to which civilization is ecologically threatened. It is probable that the earth is in the grips of a global ecocrisis. But establishing that judgment in any systematic way is beyond my scope. I take it on the basis of expert opinion, such as that of the noted conservation biologist and entomologist Edward O. Wilson, that our earth verges on a mass extinction of life from anthropogenic causes (1992, 32). I also have been persuaded by the arguments of John Firor (1990) that the prospect of global climate change threatens civilization. And recent scientific studies, such as those collected in *Biotic Interactions and Global Change*, confirm the judgments of Wilson and Firor that climate change and the extinction of species are not unrelated (Kareiva et al. 1992). More generally, discussions of the gravity of climate change, extinction of species, and other aspects of ecocrisis are ongoing in a variety of scientific journals. A recent cover of *Nature*, for example, features a photograph of the earth taken from space with the caption "Confronting Climate Change." The lead editorial in this issue asserts that climate change poses a threat not only to the comfort but potentially to the survival of our species (Maddox 1992a). Similarly, a summer 1991 issue of *Science* features biodiversity on its cover and in a large collection of articles. The lead editorial asserts that the preservation of species is more than a matter of conscience and likely involves human survival (Koshland 1991).

On the basis of authoritative opinion, then, I claim that the earth is in the grips of ecocrisis, even as I simultaneously admit that science has not resolved all the questions as to the nature of this crisis. What is it, skeptics sometimes ask, that climate models predict? How can we take action if the empirical evidence is inconclusive? And how can we claim that mass extinction is in the offing when the number of species on earth cannot be estimated even to the nearest order of magnitude? One reason I claim that life and civilization are threatened is because neither the issue of climate heating nor that of extinction of species is as ill-defined as skeptics suggest.

The concentration of atmospheric CO_2 is increasing at a historically unprecedented rate. As noted above, the primary sources of this rise are the consumption of hydrocarbons (burning fossil fuels adds CO_2 to the atmosphere) and the destruction of the tropical rainforests (through photosynthesis trees fix carbon and store it in their leaves, trunks, and roots). By developing rainforests on the Western economic model, which assumes that ecosystems are valueless until humanized, vast amounts of carbon are returned to the atmosphere. Among other indicators, none of which is decisive in its own right, the diminishing size of the polar ice caps, the bleaching of coral reefs, the northward retreat of heat-sensitive plant species, the extinctions of equatorial amphibian species, the disruption of established climatological patterns (such as rainfall), the overall twentieth-century increase in temperature, and the projected rise in sea level increase the likelihood that anthropogenic climate heating is a fact. It is also a fact that the inertia of the system that causes climate heating precludes any quick fix: change will take decades or centuries rather than years.

Similarly, and tragically, species continue to go extinct at an increasing rate, largely as a consequence of human factors. Wilson argues that the Energy-Stability-Area Theory of Biodiversity (in general terms, that size of the ecosystem, stability of climate, and available solar energy are the determinants of biodiversity) has been confirmed by observations of *predicted* diversity of life in areas of climatic stability, such as tropical rainforests (1992, 206), even while the forces of "economic progress" relentlessly destroy this habitat. "By every conceivable measure," Wilson argues, "humanity is ecologically abnormal. Our species appropriates between 20 and 40 percent of the solar energy captured in organic material by land plants. There is no way that we can draw upon the resources of the planet to such a degree without drastically reducing the state of most other species" (272).

As virtually all informed environmentalists know, neither climate change nor mass extinction is unprecedented. The questions, more than anything, are what can we learn from the past? And what should we do now? Previous climate changes, such as the neothermal shifts at the end of the last ice age, had revolutionary consequences for humans. The transformation from hunting-gathering to agriculture was precipitated in part by those events. But that process of climatological transition extended over thousands of years, allowing time for flora and fauna, as well as human culture, to adapt. The more rapid change portended by anthropogenic climate heating, on the order of decades and centuries rather than millennia and tens of thousands of years, makes it unlikely that most of the earth's flora and fauna will adapt. As noted, ecocrisis is better grasped as a web of connected variables than as a consequence of any single factor. At present, the relentless destruction of habitat, itself tied to rapidly increasing population levels, is a greater threat to biodiversity than carbon dioxide emissions.

Temporal considerations also pose a challenge to our own ability to adapt—to transform the social matrix. Science suggests a plausible guide for action: prudence is a virtue when dealing with global ecosystems. We commit the infamous appeal to ignorance by concluding that it is safe to continue doing what we are doing since we cannot absolutely prove that global climate change or mass extinction is in the offing. Proof in the strict sense exists only in the mathematical sciences. That kind of proof has not and will never be achieved for complicated (multivariate) systems evolving in real time. But empirical models predicting global warming are rapidly becoming more sophisticated and coherent as additional variables are incorporated and computational techniques are refined (Wigley and Raper 1992). Similarly, conservation biology is increasingly adept at constructing multivariate models that explain anthropogenic extinction of species (Soulé 1991).

As the Earth Summit convened in Rio de Janeiro in June 1992, the international scientific community presented a relatively united front on a variety of issues, including biodiversity, climate heating, overpopulation, and the imperative nature of timely action. Consider three cases in point:

- In August 1990 the members of the Ecological Society of America (ESA) adopted the Sustainable Biosphere Initiative (SBI). In the opinion of the members, the risk of global ecocatastrophe warranted the SBI, despite uncertainties in environmental risk assessment, climate modeling, and other issues. Establishing three research priorities—global change, biodiversity, and sustainable ecosystems—the SBI aims to provide scientific knowledge on important environmental issues for decision makers, including the public. Simon Levin, past president of the ESA, explains the rationale. "Policy makers and the public must be provided with the range of possible outcomes, with best estimates concerning probabilities, and then decide what risks [they wish] to take, and how to balance costs and benefits. At present, these *decisions are being made without proper scientific input*" (1992, 217, my emphasis).

- In November 1991 an international conference on a scientific agenda for environment and development into the twenty-first century (ASCEND 21) was held in Vienna, organized by the International Council of Scientific Unions in cooperation with the Third World Academy of Sciences. The summary document from the conference emphasizes "the central importance of the precautionary principle, according to which any disturbances of an inadequately understood system as complex as the Earth system should be avoided" (ASCEND 1991, 1). And while the conference communiqué identifies several problems as being of "the highest scientific priority," including climate change and loss of biological diversity, it also emphasizes that solutions must involve considerations of ethics and values. *Scientific solutions alone are not feasible.*

- In February 1992 the Royal Society of London and the National Academy of Science issued an unprecedented joint statement, warning that time is running out for effective response to such problems as overpopulation, climate heating, and biodiversity. Emphasizing the groups' "deep concern," the communiqué asserts that if present patterns of human activities continue, then "science and technology may not be able to prevent either irreversible degradation of the environment or continued poverty for much of the world" (Maddox 1992b). Mitigation of CO_2 emissions, energy conservation, and population control are among the report's many recommendations. And biodiversity must be protected, because extinctions are irreversible events that have "serious consequences for the human prospect in the future." The statement also underscores the new slogan for the 1990s: "Act locally, think globally." As John Maddox, editor of *Nature*, points out (1992c),

> solutions to global ecocrisis entail international agreements. Finally, warns the joint report, science and technology alone cannot solve the many challenges posed by global ecocrisis. No political, economic, or technological solutions for environmental crisis exist outside some value-laden assessment and consideration of the human condition, primarily because technical administration and, indeed, science "left without moral and political guidance" are themselves part of the etiology of ecocrisis.

Assuming, then, on the basis of the most recent scientific information that climate heating, extinction of species, and overpopulation (among other factors) are pushing our earth ever deeper into ecocrisis, and granted also that most of the salient indexes of environmental degradation are worsening rather than improving, just what religion has to do with all this remains an issue. Can it be that the environmentalists who have charged religion with culpability for ecocrisis are wrong? And even if so, how could religious discourse contribute meaningfully to solutions? Chapter 1 begins with these questions.

I

Religion in the Context of Ecocrisis

The term "ecojustice" expresses the determination to hold together the concern for justice as a norm for human relations and the awareness that the human species is part of a larger natural system whose needs must be respected. Because Christian ethical theory, if not practice, has focused on justice and largely ignored ecological reality, we Christians should be particularly eager to bring into our deliberations persons who are keenly aware of the dangerous ecological consequences of our earlier humanitarian actions and wisely warn us against heedless continuation.—John B. Cobb, Jr., *Sustainability: Economics, Ecology, and Justice*

The claim that religion has a role to play in resolving ecological crisis seems on its face naive if not patently absurd. Few seminaries and divinity schools incorporate environmentally relevant instruction into their curriculum. And if religious leaders are not ecologically informed, there seems little reason to think that religion can guide the faithful toward a socially just and sustainable society. In truth, other than a few intellectuals, literary critics, or religious believers, few of my contemporaries seem to believe that religion has any *pragmatic function* at all. Environmentalists themselves, trailing in the wake of such scholars as Lynn White, Jr., characteristically believe that Judeo-Christianity—its values, metaphysics, and institutions—is the cause of ecocrisis. Religious believers, for their part, often think of their faith as concerned more with otherworldly issues, such as salvation of the soul, than with such mundane matters as pollution or recycling. And, almost uniformly, scientific professionals presume that what environmental crisis calls for is more ecologically oriented education combined with more management—better plans and designs for using the earth's resources, closer monitoring of land use and pollution levels, and so on. Believing in the separation of church and state, ecology-minded voters think that they can elect an "environmental president" or legislators who will defend nature. If nothing else, the Department of Interior, EPA, and Department of Energy will see to conservation. And intellectuals generally, though there are many exceptions, dismiss religion as a dead letter: science is truth and religion is naught but superstition.[1]

Appearances, however, are deceiving. Consider the twenty-some years between Earth Day One and the present. In spite of environmental achievements, which I do not mean to belittle, *on the whole* conditions are worse. For example, the Ehrlichs brought the issue of population to public attention even before Earth Day 1970. The American population continues to grow; worldwide, nearly a billion and a half people have been added in twenty years. We are rapidly closing in on six billion people. One doesn't have to be a genius to know that unless we can defuse the population bomb we have little chance of resolving the ecocrisis. Population growth is, without question, tied to other facets of global ecocrisis—resource depletion, habitat destruction, loss of biodiversity, and rising levels of atmospheric carbon dioxide. Atmospheric CO_2 levels, for example, have increased from approximately 315 to 355 parts per million between 1958 and 1990—a stupefying rate of change that is driven by a growing population, use of hydrocarbons for energy, and deforestation.[2] Authoritative scientific sources argue that climate heating is no longer a hypothetical but an actual exigency of such consequence that humankind must act even in the face of uncertain knowledge. And E. O. Wilson estimates that the loss of species has increased from *a minimum* of one thousand per year during the 1970s (1984, 122) to more than ten thousand per year (1992, 351). He writes that "humanity has initiated the sixth great extinction spasm, rushing to eternity a large fraction of our fellow species in a single generation" (1992, 32).

One more example: between 1970 and 1990 the Gross National Product grew by approximately 40 percent. Over that same span the Index of Sustainable Economic Welfare (ISEW), which includes variables excluded from the national income accounts, such as costs of air and water pollution, loss of farmland, and depletion of nonrenewable natural resources, declined by an estimated 10 percent (Cobb and Daly 1989, 418–19). The very measure of short-term economic success, a statistic dear to politicians and economists, obscures the deterioration of the ecosystems necessary to sustain long-term economic productivity. How many more decades of "economic prosperity" can pass before nature is damaged beyond the point that dynamic equilibrium can be maintained and the managed society becomes a reality? Bad ecology does not make good economics.

Again we confront the paradox of environmentalism. More than two decades of almost continuous action, most of it well-intentioned, indicate a failure to stem the drift of Western culture toward ecological breakdown. Like Alice in Wonderland, the faster we run, the farther away seems our goal, minimally defined as sustainability. *Dynamic equilibrium*, a characteristic of natural ecosystems relative to cultural systems, seems ever more elusive.[3] The salient indexes—extinction of species, atmospheric CO_2 levels, population growth, ISEW—are moving in the wrong direction. One reason is a poor grasp of the nature of logarithmic growth by politicians and the lay public. Another reason is inadequate knowledge of the dynamics of ecosystems. Island biogeography, for example, has revealed flaws in the basic assumptions that govern the preservation of wildlife habitat. And the political will to address the anthropogenic causes of ecocrisis has not materialized: voters tend to believe that technology will solve the problems. Or they even believe that the problems are not real, inventions of the media or "ecofreaks"—radicals who want power to satisfy their own selfish agendas even if they undercut civilization. The so-called wise-use conservation movement attacks environmentalists as modern-day Cassandras, as proponents of "apocalyptic environmentalism" who hope to gain political advantage through manufactured crises and fear tactics.

So framed, the limited successes in cleaning polluted air and protecting endangered species (almost exclusively large mammals and a few birds and fish) conceal a potentially catastrophic outcome. History confirms that cultures often emerge, flourish for hundreds or perhaps thousands of years, and then collapse as a consequence of habitat degradation (Ponting 1992). The process, in terms of human time, is relatively slow and insidious. The earth is not degraded all at once; the sky doesn't fall. But there comes a time when environmental ruin leads to cultural breakdown.[4] Even more important for our purposes, however, is that the steady worsening of ecocrisis (as measured by relevant indexes), in spite of society's efforts to ameliorate adverse influences, indicates something more than the potential for cultural catastrophe. The paradox of environmentalism implies that *the secular standpoint*—the notion that humankind stands on the stage of history as the master and possessor of language—has its limits. The secular standpoint, epitomized by the attempt to manage planet Earth, is likely more part of the problem than any solution. Perhaps the degradation of the earth indicates that religion, whatever its role in creating ecocrisis, and contrary to the expectations of environmentalists, voters, and ecologists, among others, can play a positive role in conservation. The thesis before you, then, is that religion is a *necessary condition* for the resolution of ecocrisis[5] (a claim distinct from the premise that religion alone can solve the crisis).

A Brief History of Prejudice: Environmentalists against Religion

Most environmentalists believe that religion, especially Judeo-Christianity, thwarts rather than advances any societal effort to achieve sustainability. For almost twenty-five years I, too, believed the party line. My prejudice started with a vague suspicion, acquired by reading John Muir and Aldo Leopold, that hardened into faith upon publication of Lynn White's famous essay, "The Historical Roots of Our Ecologic Crisis" (1967). Unlike many environmentalists, however, I did not publish my opinions, in part because I saw some possibility that native American religions and possibly Zen Buddhism offered remedies and because I believed that to criticize Judeo-Christianity would be overkill. In the past few years, however, I have been led, largely by the paradox of environmentalism, to reconsider this belief. I have also learned how to read White's essay in a different way (see chapter 6, below). Whatever any environmentalist's opinion of religion, the failure of "critical reason" to resolve the ecological crisis implies the importance of reconsideration.

However, many environmentalists, in spite of evidence to the contrary, continue to think of religion as the enemy. That bias, perhaps more than anything else, frustrates collective efforts to develop more useful habits of action. By all available evidence, the environmental intelligentsia has been unable to muster the political support necessary to redirect American society toward sustainability. Many ecologists who are gravely concerned about biodiversity, overpopulation, climate heating, and other environmental issues remain part of a scientific silent majority, deeply involved with their research and little else. And the paradox of environmentalism suggests that neither scientific nor ethical arguments have caused environmentally harmful public policies and social norms to be replaced by productive ones. In other words, the environmental movement has not proven itself adequate to the task of creating a democratic consensus on sustainability. Religious discourse, if nothing else, is one possible way a democratic people might achieve solidarity—that is, create the political will to elect leaders who in turn would create public policies that lead toward sustainability.

Those environmentalists who refuse to entertain the hypothesis that religious discourse might be relevant would do well to reconsider the lessons of pragmatism. Charles Sanders Peirce, America's first pragmatist, counseled (1955) that above all we must not block the path of inquiry. Of course, no environmentalist is so simple-minded to believe that if religion were eliminated, the ecocrisis would resolve itself. Neither do the faithful think that if environmentalists were eradicated, ecological problems would disappear. In any case, conventional environmentalist wisdom has it that religion is the cause of the crisis and that religion must be either transformed or abandoned if we are to take effective action.

Consider, for example, John Livingston, the noted Canadian conservationist. He writes in *The Fallacy of Wildlife Conservation* that "there has not been any perceptible shift in any branch of traditional western religion with respect to the relationship of man and nonhuman nature" (1981, 56). The problem, according to Livingston, is that religious beliefs, such as salvation of the soul and eternal afterlife, fly in the face of scientific belief. The faithful also ignore issues of immediate consequence, such as the ecological effects of human populations on regional and global ecosystems. "I know of no formalized western theology," he continues, "that even admits the kinds of questions that are raised by ecologic insight, much less begins to face them" (58). What relevance, for example, could a biblical tradition founded more than two millennia ago have to contemporary ecological problems? A thousand saints working for a thousand years could not find one scintilla of insight in the Bible into the complexities of the global greenhouse, biodiversity, or world population levels. What use, Livingston asks, is Judeo-Christianity in a time of ecocrisis?

Prejudice is prejudgment and manifests more than anything a refusal to consider evidence contrary to a foregone conclusion. Environmentalists generally, and Livingston specifically, ignore arguments that counter their opinion of religion. In 1970, some eleven years before *The Fallacy of Wildlife Conservation* was published, for example, Francis Schaeffer argued in *Pollution and the Death of Man: The Christian View of Ecology* that Christians must be responsible stewards for creation.[6] Schaeffer is a biblical inerrantist, believing that the Bible is the literal truth—that God, for example, made the Creation in six days. *Man's* (Schaeffer's term) continued abuse of the natural world, he argues, is a desecration of the Creation. Thus Christians, made in God's image and given dominion over the earth, are obligated to care for Creation.

Livingston's obvious rejoinder would be that Schaeffer's case for Christian stewardship is not ecologically informed and therefore bound to fail. Clearly, Schaeffer cannot with self-consistency entertain the kinds of questions raised by a *scientifically based* consideration of evolutionary relations between humans and the earth. Livingston might justifiably claim that Schaeffer's thesis is contradictory to the evolutionary paradigm and hence antiscientific. Further, Livingston might argue that Schaeffer fails to make changes in his beliefs consistent with ecological science, since his basic attitude is that "man" is given dominion over creation. Perhaps, then, hard-line environmentalists can at least forestall if not defeat arguments like Schaeffer's and thereby maintain their opinion that religion is useless in a time of ecocrisis.

Other arguments, however, address the question of humankind's relation to nature from an explicitly ecological *and* Judeo-Christian perspective. In light of Rosemary Ruether's *New Women, New Earth*, published in 1975, six years before *The Fallacy of Wildlife Conservation*, Livingston would have a more difficult time defending his thesis. Yet here he might claim, taking a more aggressive posture, that Ruether did not have an *ecotheology* until 1983, when her *Sexism and God-talk* was published, so that his charge from 1981 (that no Western theology deals with questions raised by ecology) was accurate. But this defense founders on John Cobb's *Is It Too Late? A Theology of Ecology* (1972). Cobb, a preeminent theologian, admits that Judeo-Christianity is culpable, at least in part, for ecocrisis. But he argues that an ecological Christianity, one combining an ecological perspective with a biblical tradition, is not only feasible but essential to dealing with ecocrisis.[7] (Although not an explicit ecotheology, Cobb's book *The Liberation of Life*, cowritten with the biologist Charles Birch, also challenges Livingston's thesis and points to the many possibilities for the relevance of the biblical witness to environmental affairs. "The human calling," they suggest, "is to respond to Life here and now so that life on this planet may be liberated from the forces of death that now threaten it" [1981:202].)

In part Livingston, like many other environmentalists, including myself, in the late 1970s and early 1980s, was worshiping at the altar of Lynn White's influential article "The Historical Roots of Our Ecologic Crisis."[8] Here White argues that ecocrisis will continue to worsen until "we reject the Christian axiom that nature has no reason for existence save to serve man" (1973, 29).[9] Judaism (White's term) is culpable since Christianity inherited its biases from the tribes of Yahweh, biases that include the notion that "man" is made in God's image. White also charges Christianity with destroying pagan animists, who worshiped the earth and its flora and fauna. He concludes that Judeo-Christianity will have to be either eradicated or transformed for any religious solution to environmental crisis to be feasible. If Judeo-Christians decide to transform their faith, White recommends Francis of Assisi as an appropriate model. If they choose to abandon their faith, he recommends Zen Buddhism as a substitute—although he does this neither in a systematic theoretical way, by exploring the philosophical nuances of Zen, nor in a practical way, by examining the actual relations between Zen Buddhists and the nonhuman world.

John Passmore's book *Man's Responsibility for Nature* (1974) came soon after White's article. Passmore's position remains important in some ways, since he does not lay the blame for humankind's abuse of nature solely on Judeo-Christianity. He holds that Judeo-Christianity, in combination with the Scientific Revolution, provided a legitimating rationale for the modern age, the age in which Americans are still living. But like Livingston and other environmentalists, Passmore sees little chance for religion to help resolve ecocrisis, largely because he thinks that environmental ethics is incompatible with the growth dynamic of Western civilization, and hence is incompatible with a theology that has supported it.[10]

Since White wrote "The Historical Roots of Our Ecologic Crisis," indeed, since Livingston wrote *The Fallacy of Wildlife Conservation*, several things have happened. For one, the charge that Judeo-Christianity caused the environmental crisis has been discredited as an oversimplification of a complicated skein of historical events. Granted, Judeo-Christian attitudes and beliefs are relevant to explaining environmental problems. But no capable inquiry abstracts Judeo-Christianity from the greater whole of history as an explanatory variable. Apart from the course of Western civilization, including the Renaissance and the Reformation, the Scientific Revolution and the Enlightenment, as well as the geniuses who envisioned the modern age, such as Francis Bacon and Adam Smith, no discussion of Judeo-Christianity and its influence makes sense. Further, other cultures, even aboriginal cultures, have been capable of assailing nature. The American southwest, for example, is dotted with sites abandoned by prehistoric aborigines after they drove the habitat beyond dynamic equilibrium.

Within the Judeo-Christian tradition itself (conceived in an inclusive way that does not marginalize any denomination) there has been growth in interest among the faithful—theologians, ministers, educators, and lay people—in the environmental crisis. In the past decade an array of theological material that explicitly embraces ecology has been published and disseminated in religious circles. A recent book by Joseph Sheldon, *Rediscovery of Creation: A Bibliographic Study of the Church's Response to the Environmental Crisis*, lists 1,700 references.[11] Many writers have argued that Judeo-Christians do have a responsibility to treat the natural world in an ethical manner.[12] The upshot is that ten years after Livingston's *Fallacy of Wildlife Conservation*, no one can claim that no Western theology entertains questions raised by an ecologically informed understanding of humankind's relation to nature. Not only have an increasing number of religious thinkers addressed the ecological crisis formally, but they have started to address specific problems like biodiversity and overpopulation. Jay McDaniel's *Of God and Pelicans: A Theology of Reverence for Life* (1989) exemplifies the possibilities for an ecologically informed theology, and Susan Bratton's *Six Billion and More: Human Population Regulation and Christian Ethics* (1992) is a treatment of the social, ecological, economic, and spiritual dimensions of overpopulation. These are just two among hundreds if not thousands of relevant titles. The ecotheological literature is so large and growing so rapidly that no reader can hope to stay abreast of it.

Of course, the existence of this literature and the familiarity of a few intellectuals with it is one thing. Changing ecologically significant attitudes and practices among religious people is another. As mentioned, divinity schools do not usually teach ecotheology or environmental ethics. Neither do lay people usually examine the ecological implications of their faith. But the potential for the faithful to respond to ecocrisis, consistent with their faith traditions and claims to ultimate knowledge, is real. And it is crucial to our future. In an important edited collection, *Religion and Environmental Crisis*, Eugene Hargrove asserts that the debate initiated by Lynn White has gone nowhere, "beginning and ending inconclusively with the same set of issues" (1986, xvi). Hargrove suggests that merely finding Judeo-Christianity accountable for ecocrisis or observing that the biblical tradition has not yet effectively resolved the ecological crisis is self-defeating. The important issue facing us, he contends, is to find "ways for major religions to respond to the environmental crisis" (xvii). Hargrove recommends taking a Wittgensteinian path—one I will follow, beginning with my sociolinguistic perspective—from where we are, mired in an ever worsening environmental crisis, to where we might hope to be—namely, living in a society that, however far from sustainability, has reversed the direction in which the indexes of environmental degradation are moving.

Religion from a Sociolinguistic Perspective

The notion that humans are usefully described as storytelling culture-dwellers frames a nonpartisan inquiry into religion but does not preclude alternative descriptions of human beings, either religious, such as the description of "man as created *imago Dei*," or scientific, such as *Homo sapiens*. From a sociolinguistic perspective the notion that "man is made in the image of God" is no more and no less than a culturally dominant definition. Where the sociolinguist differs with the religious believer is in recognizing the possibility of alternative definitions; no one definition is intrinsically favored over any other. Which is to say that *Caring for Creation* takes a sociolinguistic stance, a posture that situates religious discourse in relation to environmental crisis. I claim only that it is useful to think of religion from such a perspective, not that this is the right way to think of it. Richard Rorty suggests that "viewing inquiry as recontextualization makes it impossible to take seriously the notion of some contexts being intrinsically privileged, as opposed to being useful for some particular purpose" (1991b, 110).

Religious believers who make claims to ultimate knowledge will not be entirely comfortable with my approach. They cannot in fact take such a perspective, for to do so requires suspending the conviction they have in their claims to ultimate knowledge. For example, religious conservatives and Native Americans believe in a self-sustaining spiritual reality that exists apart from human discourse. (See chapter 4, below.) For such believers a sociolinguistic perspective is germane neither to the belief in nor to the justification for their faith. Fortunately, my thesis does not require that believers abandon their claims to ultimate knowledge in order to care for creation. Only the sociolinguist needs to suspend ultimate belief, even while affirming the importance of specific claims, since they are of great consequence to the faithful. Indeed, without the faith traditions to which I appeal, there would be no possibility of making my case that religion can lead us to care for creation. My argument, however, has nothing to do with either the adequacies or the inadequacies of different metaphysical convictions, that is, rival claims to ultimate knowledge. I aim to show only that a democratic consensus on caring for creation can be reached from metaphysically diverse and sometimes antagonistic positions of faith.

In the context of sociolinguistics, religious discourse is grasped as an important strand in the web of belief—the legitimating narratives that define a culture and guide human behavior. George Lindbeck (1984), a leading advocate of a sociolinguistic (or sociocultural) perspective on religion, elaborates.

> A religion can be viewed as a kind of cultural and/or linguistic framework or medium that shapes the entirety of life and thought. . . . It is similar to an idiom that makes possible the description of realities, the formulation of beliefs, and the experiencing of inner attitudes, feelings, and sentiments. Like a culture or language, it is a communal phenomenon that shapes the subjectivities of individuals rather than being primarily a manifestation of those subjectivities. It comprises a vocabulary of discursive and nondiscursive symbols together with a distinctive logic or grammar in terms of which this vocabulary can be meaningfully deployed. Lastly, just as a language (or "language game," to use Wittgenstein's phrase) is correlated with a form of life, and just as culture has both cognitive and behavioral dimensions, so it is also in the case of a religious tradition. Its doctrines, cosmic stories or myths, and ethical directives are integrally related to the rituals it practices, the sentiments or experiences it evokes, the actions it recommends, and the institutional forms it develops. [33]

Such a perspective should not be perceived as the one true outlook on religion but only as a useful one, one that informs us about religion's role in a time of ecocrisis and, more generally, helps reconstruct the web of belief. If nothing else, a sociolinguistic viewpoint might help believers place their faith in the context of alternative traditions. More specifically, Lindbeck's approach is useful in several ways.

First, it opens the path of inquiry, since no one doctrine is initially privileged over any other. All creeds can be appreciated as "ways of life" that determine human behavior. Religious believers often privilege their discourse to the exclusion of others: their legitimating narratives become final vocabularies used to judge others as meaningless. Viewed sociolinguistically, this privileging is metaphysical, that is, removed or isolated from the open-ended play of language so that absolutes (so-called transcendental signifieds) are created. My approach recognizes the metaphysical claims inherent in religion without privileging any one creed. Clearly, religions differ in their potential for making adaptive responses to ecocrisis. My goal, however, is not to find the "best" (or greenest) religion but to explore the different potentials that exist across the spectrum of faith. Insofar as the claims to ultimate knowledge made by the faithful hinder rather than advance efforts to achieve solidarity, my viewpoint is useful in showing the possibility of a democratic convergence on the importance of sustainability that also remains consistent with religious diversity. Any attempt to resolve competing claims to ultimate knowledge involve metaphysical and theological arguments that are beyond the scope of my inquiry.

Second, it is consonant with the reality of time. Religious doctrines, understood as paradigmatic, are thus open to the possibility of change, just as are the claims to truth of physics or biology. As Lindbeck puts it, innovation is comprehended "not as proceeding from new [religious] experiences, but as resulting from the interaction of a cultural-linguistic system with changing situations. Religious traditions are not transformed, abandoned, or replaced because of an upwelling of new or different ways of feeling about the self, world, or God, but because a religious interpretive scheme (embodied . . . in religious practice and belief) develops anomalies in its application in new contexts" (39). Lindbeck implies that even the great prophets and charismatic leaders (so-called world historic individuals), such as John the Baptist and Jesus Christ, respond to perceived inadequacies of religious tradition and its institutionalization. More crucially, a sociolinguistic perspective enables us to grasp the prophetic and critical traditions of Judeo-Christianity as a possibility in the here and now, as the faithful begin to engage the reality of ecocrisis. Contrary to the thesis that Judeo-Christianity must either be abandoned or replaced, Lindbeck's approach emphasizes the potential of religious discourse to provide meaningful guidance in a time of ecocrisis (changing circumstances) while remaining consistent with tradition. As I shall argue in chapter 4, the entire spectrum of faith, from conservative to radical, can address ecocrisis in a way that is intellectually persuasive and emotionally convincing.

Last, and perhaps most important, a sociolinguistic perspective helps establish a balance between religion and science generally and between religion and ecology more specifically. As mentioned, part of environmentalists' prejudice against religion stems from their belief that religion is antithetical to ecology, exemplifying the traditional opposition between faith and reason. From a sociolinguistic perspective, such an opposition cannot be maintained, since religious discourse is as important as ecological discourse in moving our culture toward sustainability. Religion and science, in other words, are more usefully viewed as collaborative than as antagonistic. Both are necessary conditions for the resolution of ecocrisis in a democratic society. Such an idea, however, is so important and iconoclastic that elaboration is worthwhile.

The Collaboration of Religious and Ecological Discourse

Some of the most distinguished scientists and philosophers of the twentieth century, Albert Einstein, Erwin Schrödinger, Ilya Prigogine, Richard Rorty, and T. S. Kuhn, question any complete distinction between science and the cultural context within which it exists. The traditional (positivistic) picture of science as providing a universally true account of the world, good for all people in all places and times, accurately describes how scientists in particular have described their claims to knowledge. But the idea that scientific statements are expressed in a language that the universe prefers, one that "cuts things at the joints," is according to Rorty, "a pretty conceit" (1991b, 80). From my viewpoint the fact that *science enables prediction*, regardless of additional claims to ultimate knowledge, justifies its methods, truth claims, and utility.

The works of T. S. Kuhn (1970) and others, such as Jean-François Lyotard (1984), are useful descriptions of science. Kuhn argues that science has not only a historical but a social dimension. Modern science is an artifact of history, the product of certain individuals with certain kinds of motives acting in context. Contrary to the idea that some "reality" beyond the human world unconditionally determines truth, the inescapable historical and sociological reality is that the truth claims of science are empirically underdetermined. Ilya Prigogine, a Nobel laureate in physics, notes that many descriptions are possible for any natural event and that these may be either complementary or opposed. Although "they all deal with the same reality," Prigogine continues, "it is impossible to reduce them to one single description. The irreducible plurality of perspectives on the same reality expresses the impossibility of a divine [or objective, universal, and eternal] point of view from which the whole of reality is visible" (1984, 225).

Not only is no single description definitive, being simply one among many, but even the so-called hardness of scientific fact is a consequence of language. This does not imply that the process of scientific inquiry is founded on whimsy and fantasy. The claims that scientists make presuppose rigorous standards, including objectivity. The recent refutation of the claims by some scientists to have discovered "cold fusion" is an example of science at work. But science provides no absolute perspective on space-time, so that its true propositions mirror reality—that is, reflect it in a one-to-one isomorphic relation. A representationalist theory of knowledge is an erroneous description of science and the kinds of activities that scientists carry on. But the demise of representationalism does not imply that claims to scientific truth are without epistemic justification. The objectivity of claims of scientific truth comes from several sources, especially the standards of a community of researchers operating within a shared framework of presuppositions, the relentless effort to falsify conjectures by putting them to empirical test, and the open-endedness of the process of inquiry.

Some readers may think that this description of science is too relativistic, that it undermines the claims made in the Introduction about global climate change and the extinction of species. But the truth value of these statements does not depend on any claim that they "mirror reality." One reason is that claims to ultimate knowledge tacitly presuppose the existence of a community of scientific researchers and their procedures. From inside language—that is, from a position that recognizes the linguistic practices of scientists—the actual descriptions that state facts exhaust the possibility of true statements about reality. Claims to ultimate knowledge go beyond scientific truth, which is to say that they are metaphysical. True statements presuppose a process of scientific inquiry that warrants their probability—that is, the likelihood that they are accurate. There is no sense in saying that true statements are the "mirror of nature." A sociolinguistic perspective, then, is not relativistic but realistic insofar as it ties an account of truth to the existing practices of scientific communities. And in a time of ecocrisis these practices are essential to the future: scientific truth is useful; we cannot do without it.

The sociological and historical reality is that scientific knowledge is better described as a process of ongoing inquiry that seeks truth rather than as a window through which humans glimpse "the real world." Scientists are not priests who put us in touch with some supra-historical reality, the way things really are, although some scientists make that claim. This does not mean that there are not things that exist independently of human beings; neither does it mean that scientists are investigating a dreamworld or fantasyland that they just make up. "To say that truth is not out there," Rorty claims, "is simply to say that where there are no sentences there is no truth, that sentences are elements of human languages, and that human languages are human creations" (1989, 5). So construed, true statements state facts about the world, the things in the world, and the relations among the things in the world. True statements also have predictive utility, a pragmatic but not insignificant justification for the cultural value of science; and they are grounded in the scientific tradition of open-ended inquiry—a model for conversation that should inspire us all.

The historians, sociologists, and philosophers of science have not alone established this account of science. Einstein, though he hung on to the notion of Universal Truth, also grasped the cultural contingency of science—that is, the fact that science always exists in a social context and necessarily finds its justification within that context. Science, he writes, "provides us with powerful instruments for the achievement of certain ends, but the ultimate goal itself and the longing to reach it must come from another source. . . . The knowledge of truth as such is wonderful, but it is so little capable of acting as a guide that it cannot prove even the justification and the value of the aspiration toward that very knowledge of truth. Here we face, therefore, the limits of the purely rational conception of our existence" (1954, 42).

Similarly, Schrödinger, who shared the 1934 Nobel prize in physics with Paul Dirac, provides a useful perspective. Compared to Einstein, Schrödinger admits to an even more intimate relation between the quest for scientific truth and culture. He notes that "there is a tendency to forget that all science is bound up with human culture in general, and that scientific findings, even those which at the moment appear the most advanced and esoteric and difficult to grasp, are meaningless outside their cultural context." Scientific practitioners themselves, Schrödinger fears, too often neglect this fact. The consequences of such oversight, he warns, can be devastating. "A theoretical science . . . where this is forgotten, and where the initiated continue musing to each other in terms that are, at best, understood by a small group of close fellow travellers, will necessarily be cut off from the rest of cultural mankind; in the long run it is bound to atrophy and ossify however virulently esoteric chat may continue within its joyfully isolated groups of experts" (1952, 109–10).

Einstein and Schrödinger and many others, then, help us realize that science is not a self-contained and self-justifying absolute, since it exists within a cultural context. And this cultural context is essential in providing any descriptive account of both the existence of a community of scientific researchers and their methods. In sum, science and culture are deeply entwined. But this revelation is startling, since it implies that the perceived opposition between science and religion is more a reflection of the different kinds of statements they make and linguistic traditions they represent than an indication that they are absolutely at odds. Further, the sociolinguistic redescription of science changes our conception of the presumed *metaphysical divide* between facts and values. Both the scientist and the religious ethicist traditionally view science as dealing with the facts of the world—what is actually the case—and ethics as dealing with values, with an ideal world—what ought to be the case. The split between facts and values has been institutionalized within the so-called two cultures. But the sociolinguistic perspective permits at least a possibility of placing science and religion in a collaborative rather than antagonistic context. This would not collapse facts and values into each other. The point is that both science and religion are relevant to the problem of finding solutions for ecocrisis.

Such collaboration is difficult to describe from either a religious or a scientific standpoint—both presuppose that there is but one true account of reality, and for that reason alone a sociolinguistic perspective is useful. For a believer who accepts the doctrine of biblical inerrancy, for example, there are many outright contradictions between the truth claims of religion and science. The creation story in Genesis, for example, is read as being in opposition with the evolutionary account; either one or the other but not both is true. Consequently, those believers who wish to retain their biblical beliefs advance the themes of "creation science" and attempt to prove that evolutionary theory cannot be sustained (that it's inconsistent, that it cannot account for all the facts) and that the biblical account is correct. Similarly, a scientist who believes that science presents ultimate knowledge sees an outright opposition between religion and science. The creation story in Genesis is read as being in opposition with evolutionary theory. Either one or the other but not both is true. The account in Genesis can then be attacked as, for example, being incompatible with scientific facts, such as the age of the universe or the fossil record.

As I have mentioned, claims to ultimate knowledge are important to those who make them, and scientific materialists and Judeo-Christians sometimes make very different statements about the way things really are (cosmological models). But there is no reason for me to enter the dispute between scientists and religionists. In chapter 4, in fact, examples from across the spectrum of religion will show that, whatever differences exist on the level of claims to ultimate knowledge, as a practical matter religious believers employ science continually, including Christians like Francis Schaeffer, whose cosmological model contradicts that of scientific materialism. Here I wish to examine the writings of E. O. Wilson, a scientific materialist whose fact claims about the extinction of species were mentioned in the Introduction. On my reading, Wilson confirms the possibility of recontextualizing science and religion.

Wilson provides the kind of authoritative information about biodiversity apart from which no conceivable solution to the problem of the extinction of species exists. Yet he also recognizes (1984, 1992) that scientific facts are insufficient to resolve the issue, as the paradox of environmentalism implies. "Environmental problems," Wilson argues, "are innately ethical. They require vision reaching simultaneously into the short and long reaches of time" (1992, 312). Part of the problem is that science alone does not usually impel people to choose one thing over another, even though scientific information is often relevant to making a choice. One tacit question running through Wilson's newest book, *The Diversity of Life*, is that of motivation. What is it, Wilson wonders, that might move people to act on the factual knowledge that we are destroying the diversity of life on earth? In this book and elsewhere Wilson offers several possibilities, such as the appeal to biophilia (the love of life), the idea of wilderness, and (professional) environmental ethics to provide that motivation. But of all the potential sources he identifies, the one most likely to make a difference is religion. Let me repeat this important point: E. O. Wilson legitimates the judgment that religion, more than any other agent, might provide the ethical context to protect biodiversity, though he does not make that claim so explicitly.

My interpretation of Wilson's position flows out of his work on sociobiology ([1975] 1980) and human nature (1978), as well as his collaborative work with Charles Lumsden on co-evolution (1981, 1983). To oversimplify (since I completely ignore the natural history of the human species), the basic argument is that the human capacity to discriminate and make judgments of good and bad, right and wrong, has been fashioned in the crucible of evolution and remains tied with subcortical, limbic structures of the brain ([1975] 1980, 3–4). Religion is one of the ways, probably the primary way, in which human beings have expressed their distinctive human sensibilities and capacities (1978, 169ff.). Religious belief, Wilson writes, "is the most complex and powerful force in the human mind and in all probability an ineradicable part of human nature" (1978, 169). Religion reflects the deep structure of the human animal in three ways. The first level he terms *ecclesiastic*, that is, "rituals and conventions are chosen by religious leaders for their emotional impact under contemporary social conditions" (1978, 176). Which is to say that religious myth, ritual, and symbols move people at a primary, affective, or emotional level of human beingness, the level that Wilson claims "makes us people and not computers" (1992, 348). The second level of selection Wilson terms *ecological*, that is, however emotionally powerful, religion "must eventually be tested by the demands of the environment. If religions weaken their societies during warfare, encourage the destruction of the environment, shorten lives, or interfere with procreation they will, regardless of their short-term emotional benefits, initiate their own decline" (1978, 176–77). The third level is *co-evolution*, that is, the reciprocal influence of memetic and genetic evolution on each other. For a scientific materialist like Wilson, this means that religion has actually shaped human evolution. For example, "Religious practices that consistently enhance survival and procreation of the practitioners will propagate the physiological controls that favor acquisition of the practices during single life-times" (1978, 177).

Wilson's statements about religion are propositions that he claims are true, and they underscore the vital role that religion not only has played but has yet to play in human affairs. On my reading, Wilson scientifically defends the idea that religious belief which endures is adaptive, that is, it plays a positive role in survival. In the context of my argument, Wilson tears down the barrier between science and religion; each remains distinct in tradition and in claims to ultimate knowledge. But neither alone can ensure the resolution of eco-crisis. "The stewardship of environment is a domain on the near side of metaphysics," he writes, "where *all reflective persons* can surely find a common ground. For what, in the final analysis, is morality but the *command of conscience* seasoned by a *rational examination of consequences?*" (1992, 351, my emphasis). But what is religion, in this sense, except the command of conscience? And what is rational examination but science?

Clearly, *given* Wilson's emphasis on the importance of developing an environmental ethic (two of his books, *Biophilia* and *The Diversity of Life* close with chapters on the subject), for he can envision no possibility of acting to resolve ecocrisis apart from ethics, *and given* his belief that "human advance is determined not by reason alone but by emotions peculiar to our species"(1992, 348), *in addition to his stated position* that religions which endure have a positive survival or ecological value, then it follows that religion is relevant to the determination of an environmental ethic.[13] The United States remains a society of religious believers. This fact alone gives me reason to think that Americans may yet come to care for creation and politically empower a movement toward sustainability.

To go beyond this discussion is entirely feasible (for example, via either quantum theory or a coevolutionary paradigm) but not useful in this context. We have good reason, then, to put the split of science and religion in a new setting, since the metaphysical bifurcation of fact and value is no longer useful. The problem with maintaining this divide is that "it is contrived precisely to blur the fact that alternative descriptions are possible in addition to those offered by the results of normal [e.g., classical scientific] inquiries. It suggests that once 'all the facts are in' nothing remains except 'noncognitive' adoption of an attitude. . . . It disguises the fact that to use one set of true sentences to describe ourselves is already to choose an attitude toward ourselves, whereas to use another set of true sentences is to adopt a contrary attitude" (Rorty 1979, 363–64). Roy Rappaport's anthropological masterpiece *Ecology, Meaning, and Religion* (1979), supports these ideas. As Wilson notes, Rappaport's work is "an especially important contribution to the sociobiology of religion" (1978, 247). In the present context, it helps me argue that religion and science can be collaborative rather than antagonistic forms of discourse.

Rappaport's work rests on a careful distinction between "cognized models," the vernacular descriptions of *the natural world* offered by the aboriginal horticulturists he studied in New Guinea (the Maring), and "operational models," the scientific descriptions of the same *ecosystem* made by ecologists. The vernacular descriptions of the Maring, from the standpoint of positive science, are mythic, supernaturalistic, not grounded in ecological reality. To scientifically trained observers the sustainability of Maring culture in a fragile tropical ecosystem is irrational, defying explanation. How could "primitive" belief systems be adaptive, since they are fictions, supernaturalistic?

Biological scientists, especially, should experience cognitive dissonance at this paradox. There are, however, cogent explanations of the seeming paradox that "primitives," believing only in their myths, can flourish and survive in delicate tropical ecosystems. There is no reason to think, Rappaport continues, that

> the representations of nature provided us by science are more adaptive than those images of the world, inhabited by spirits whom men respect, that guide the actions of the Maring and other "primitives." To drape nature in supernatural veils may be to provide her with some protection against human folly and extravagance. . . . Because knowledge can never replace respect as a guiding principle in our ecosystemic relations, it is adaptive for cognized models to engender respect for that which is unknown, unpredictable, and uncontrollable, as well as for them to codify empirical knowledge. It may be that the most appropriate cognized models, that is, those from which adaptive behavior follows, are not those that simply represent ecosystemic relations in objectively "correct," material terms, but those that invest them with significance and value beyond themselves. [1979, 100–101]

Of course, we are not horticulturists living in the rainforests of New Guinea. But Rappaport's study has important implications. As Wilson notes (1978, 180), Rappaport's work affirms that religion can be an advantage in survival. There is no a priori reason why vernacular models and scientific models cannot inform each other: as mentioned, science does not exist in isolation from culture, and any science that believes this runs the risk of becoming irrelevant. Better, perhaps, to think of science as part of the web of cultural discourse. Science is a useful tool for religious believers and environmentalists alike. In a time of ecocrisis we need scientific knowledge. But we also need faith. The role that religion plays in human affairs, especially the importance of religion in setting a cultural agenda and, perforce, directing science, has been insufficiently appreciated.

Post-positivism, scientific and religious discourse can be described as interweaving, jointly fashioning a culturally dominant web of belief. The scientific luminaries of the twentieth century contend that the scientific quest for truth cannot exist outside a culture that gives it a reason-to-be. The institution of science is value-laden, reflecting the belief that objective inquiry yields useful knowledge for the community, functional knowledge that has survival value. A sociolinguistic perspective is sometimes misconstrued as implicitly denying the importance of scientific truths, such as John Firor's explanation of holes in the stratospheric ozone or E. O. Wilson's explanation of the extinction of species. These are "true statements about the way things are." They are not usefully conceived as "the way things are"—that is, representations of ultimate reality. Sociolinguistically recontextualized, scientific knowledge represents the highly informed and scrutinized truth claims of scientists who conduct inquiry within an established and ongoing research tradition (that continues to refine the standards of inquiry). The value of true statements is independent of any argument over their metaphysical status. Truth, in the final analysis, must be adaptive; otherwise it is self-defeating.

Science and Religion Post-Positivism

Examination of the language of science and religion reveals a common origin: ordinary or natural language. As Earl MacCormac argues, science and religion "depend upon metaphor as the linguistic device necessary to suggest new meanings. Scientists can talk to theologians and theologians to scientists, but how fruitful the conversation will be depends largely upon the relationship of the two fields" (1976, 136). As stated, *my argument is that a new metaphor—caring for creation—can engender a psychologically satisfying (emotionally evocative, powerful), religiously distinctive, and scientifically plausible ethic for our time.* Further, this metaphor has a greater promise for leading to political action than any alternative: discourse cannot lead to transformation unless it is heard. Since the metaphor of caring for creation is consistent with the Great Code (the Judeo-Christian tradition, without which the West is incomprehensible), it has an enormous potential to be both intellectually persuasive and psychologically potent.

Scientific and religious discourse are more closely linked than scientists and religious believers generally think; but ecological and religious discourse do not collapse into each other. Useful distinctions between them exist. As Loyal Rue argues in *Amythia: Crisis in the Natural History of Western Culture*, science today is perceived as undermining the root metaphor of Judeo-Christianity—God as a person. Still, Rue contends, there is no intrinsic antipathy between science and Judeo-Christianity, since evolutionary science provides a new root metaphor that can revitalize the church. I hasten to add that my objectives in this book are more limited than Rue's, since he proposes a new root metaphor for Judeo-Christianity, replacing the idea of God as a person with the idea of God as an evolutionary process. The metaphor of caring for creation leaves intact privileged metaphysical claims (for example, that God is a person or that God is evolution) while pointing to a common ground across traditions of faith in "caring for creation." *Indeed, the metaphor of caring for creation necessarily carries different meanings within different faith traditions.* The metaphor of caring for creation also crosses the boundaries between science and religion. The power of metaphor, as Eva Kittay (1987) and others argue, is that it allows mutual relations to be established between apparently separate semantic fields.

Consider, for example, the concept of "pollution." Can it be cogently defined apart from a cultural web of belief that goes beyond the strict confines of scientific discourse? Or does pollution involve value judgments—that is, cultural norms that frame physical measurements? On the surface, "pollution" seems straightforward. The dictionary definition appears uncontroversial: "The introduction of harmful substances or products into the environment: air pollution." Etymology supports the notion that pollution is an unproblematic concept. The Indo-European root is *leu*, to dirty, as in the Latin *lutum*, or mud. And certain pollutants, such as radioactive substances or greenhouse gases like carbon dioxide, seem intrinsically harmful. It follows, apparently, that the environment becomes "polluted," impure, unclean, contaminated when these materials are introduced into nature. Scientists may simply determine within a range of probability what the safe limits are, and culture can go about its business. But what is clean and what is dirty?[14] Do such constructs exist independent of cultural contexts, apart from the web of language that guides life and thought? Can science alone define pollution?

Surely the carbon dioxide that is so much involved in climate heating is a "pollutant." Yet this concept is less simple than it seems, and scientific judgment alone presents a mixed picture, one that does not resolve the issue. Biology confirms that CO_2 is required for life. It is an inevitable byproduct of the metabolism of plants (plants respire CO_2 in the dark) and animals. And life on earth is carbon-based—it is tied to nature's carbon economy (beginning some fifteen billion years ago: carbon itself is a product of stellar evolution).[15] So described, carbon dioxide is a good thing, part of life as we know it. Science also provides comparisons between different amounts of atmospheric carbon dioxide, between current levels and those existing last year, last decade, or even tens of thousands of years ago. Present measures of atmospheric CO_2, extrapolated into the short-term and intermediate future, suggest that climate heating threatens both human culture and untold numbers of plant and animal species. Apparently carbon dioxide is something both good, that is, it is necessary for life, and bad, that is, too much of it plays havoc with global ecology.

The question about carbon dioxide, however, is not the exception but the rule. Nature itself produces substances that, in sufficient concentration, are biologically lethal. For example, mercury is a deadly poison that continually leaches from rocks, flowing into rivulets and streams and rivers, ultimately to the ocean. According to Giddings, one drop of seawater contains fifteen billion atoms of mercury (1973, 23). Yet fifteen billion atoms of mercury, among trillions of other atoms in the same drop, are relatively innocuous. More generally, judgments about pollution entail questions of limits, durations of exposure, degrees of concentration, and so on.[16] For example, copper, which unlike mercury is essential for life, will in high concentrations become a toxicant. Any discussion of pollution must also consider the difference between natural and anthropogenic pollutants. Working over several tens of thousands or even millions of years, natural processes exist in relative balance. Limits are established over the scale of biological time and successful ecosystems function with those limits. With human beings the situation changes dramatically, if for no other reason than the operation of industrialized economies in oblivion to limits.

A further problem in dealing with pollution is that no absolute scientific prediction can be made of the future state of any "real-time" system, since the number of variables overwhelms abilities to compute deterministic outcomes (in contrast to the case in experimental laboratories, where other conditions can be held equal). Further, natural processes have some resistance to variance from, for example, climatological norms, but the limit of natural resiliency is difficult to predict, even in ideal cases—that is, unhumanized ecosystems. Chaos theory, among other scientific developments, offers a new language through which nature is viewed as a dynamic system so complex as to defy definitive scientific description and theoretical elucidation.[17] Novelty, or what physicist David Bohm calls "the qualitative infinity of nature," is a feature of natural process (1957, 152). Still, science can, with varying degrees of probability, predict future states of real-time systems. Yet because science deals with probabilities rather than certainties, there is necessarily a margin of doubt in bringing scientific knowledge to bear on considerations of appropriate human behavior.

Little wonder, then, that Western culture seems incapable of action, not only because arguments against the reality of global warming exist but also because there is no normative rationale adequate to move culture toward the reduction of atmospheric CO_2. As discussed, "the facts" of pollution are not as simple as they seem, since "pollution" takes us beyond *the measure* of physical parameters and the consequences for living systems to *a moral dimension* involving the meaning of life. Let me restate this important point: scientists can provide all kinds of statistical measures and quantitative information, like the amount of CO_2 in the atmosphere or the number of species that are being destroyed. The pivotal issue is determining an appropriate response.

Insofar as utilitarian individualism remains the first language of the West, then atmospheric carbon dioxide is not a pollutant in the sense of dirty, unclean, and morally repugnant. The paradox of environmentalism, in other words, or at least that part of it involved with our changing atmosphere, reflects the ideological frame into which the scientific facts are set. Climate heating is simply the price paid for the good life, an inevitable by-product of the hydrocarbon economy. Perhaps this is why Ronald Reagan said he didn't trust air he couldn't see or smell. In his calculus of value, air that was visibly polluted or had a detectable odor was good, since it had the look or smell of success to it.

Our collective inability to take effective action on climate heating confirms a lack of solidarity on how significant scientific information is. The status quo, in short, is normative and overwhelms the voice of dissent—the minority opinion that the buildup of atmospheric CO_2 is morally reprehensible, a threat to life on earth. And so on through most permutations of the environmental crisis. It is not scientific information that has been lacking so much as the will to fashion this information into a democratically determined environmental agenda. Not having any collective sense of our preferences, environmental policy has been determined by default, by the socially dominant language of utilitarian individualism. In this scenario nature is no more than a standing reserve to fuel the good life—the unlimited production and consumption of material goods. Voices of dissent, even when supported by an avalanche of scientific data, have been pushed to the periphery of the social decision-making process.

In fact, as Neil Evernden points out, the measure or "instance of physical pollution serves only as the means of persuasion, a staging ground for the underlying debate," for alternative narratives (1992b, 5). Environmental science (in any of its many iterations) is plastic, "an institutional shaman that can be induced to pronounce natural whatever we wish to espouse" (1992b, 15). Scientific judgment is often tailored to fit the modern narrative, becoming primarily an instrument for the control of nature. Alternatively stated, scientific judgment is framed by the progressive conservation movement: resourcism.[18] For insiders, who accept the modern story as unconditional, the goal of environmental science is to manage planet Earth (for example, see Clark 1989). Such experts acclaim the progress that has been made in controlling pollution, thereby advancing the good life. Science helps to confirm, in effect, that our future is secure. For example, air quality standards now exist where none were in place. Measurement confirms progress. Continued progress (which presumably confirms that the modern paradigm is not morally flawed) depends on passing new laws as needed and on the development of ecotechnologies that either lessen or forestall the buildup of pollutants.

In sum, I am arguing that both religious and scientific discourse are necessary conditions for the resolution of ecocrisis. Religious discourse offers legitimating narratives that remain outside the framework of utilitarian individualism and might also lead to normative judgments of the pure and impure, which in turn would lead toward sustainability. Post-positivism, there is no intrinsic conflict between science and religion below the level of cosmological models. From a sociolinguistic perspective the question about scientific and religious discourse is one of collaboration. Can science and religion collaborate to fashion a viable and adaptive cultural narrative? From a sociolinguistic perspective the answer is yes. From my position—inside language—metaphysical distinctions between science and religion or between facts (lumps) and values (texts) are not useful. Vernacular models—religious discourse—are required to contextualize scientific discourse in ways that lead toward sustainability. Recognition of the influence of cultural context on science might resolve the paradox of environmentalism. At present the modern narrative of utilitarian individualism rules our collective lives. Nature is no more than a standing reserve awaiting human appropriation.

Environmentalism Today

The language of experts, some of whom believe that we can manage and others who believe that we can theorize our way out of environmental crisis, dominates environmentalism today.[19] Environmental ethicists and philosophers exemplify the theorizers. They like to claim that their discourse is the way to resolve ecocrisis. There is good reason, however, to doubt that environmental ethicists per se will help large numbers of Americans redirect their behavior and make political choices. In part this is because, almost by definition, environmental philosophers are experts—intellectuals that nobody wants. The germinal twentieth-century work of ecophilosophy, Aldo Leopold's *Sand County Almanac*, was published in 1949. It remains influential not only among philosophers but also among biologists and ecologists. Yet the paradox of environmentalism calls into question any sanguine pronouncement about the efficacy of Leopold's land ethic specifically or environmental ethics generally. Environmental ethicists might do well to consider that types of language other than philosophical discourse can provide alternatives to utilitarian individualism and that any one mainstream religious denomination alone influences the normative choices of more people than all ecophilosophies put together.

Environmental philosophy has, to its credit, generated rebuttals from within its own ranks to the claim that it has solutions. The first and most notorious is that of John Passmore, who argues that since environmental ethics is effectively antithetical to the prevailing paradigm of Western culture there is little or no chance that it will be successful. More recently, Jim Cheney (1989) has argued that environmental ethics is part of the problem, part of the arrogance of humanism that keeps humankind apart from the earth, living in ecologically disastrous ways. Similarly, Rorty (1989) debunks any claim that privileges philosophical discourse. Eugene Hargrove, editor of *Environmental Ethics*, the most important journal of environmental ethics, suggests that whereas "philosophy may be better at analyzing ethical behavior . . . , religion is much more effective at changing loyalties, affections, and convictions" (1986, xiii).

A second and more influential group than environmental ethicists are the expert managers who control the institutional apparatus of the progressive conservation movement and the associated intellectual elite who maintain and refine the ideology of resourcism. This group contends that expert knowledge and technology are the solutions for ecocrisis. Efficiency, defined as the rational application of knowledge to socially legitimated goals, is the watchword of the bureaucratic mentality, argues Max Weber (1973). Such a position entails the idea that the human species can manage planet Earth. In the bureaucratic scenario the industrial growth paradigm remains supreme; as a result, no political decisions (public policy) that entail either structural or institutional change or significant changes in individual behavior are necessary. Neither does environmental crisis pose any ethical question: environmental policy administration is simply the application of knowledge to achieve the established ends of the industrial state. The environmental bureaucrat simply "deploys techniques to rationalize nature and to render it predictable, to replace its self-sustaining, 'wild' state with well-managed industrial, commercial, residential, and recreational sites" (Bennett 1987, 47).[20]

The strategy of managing planet Earth has many problems, not least of which is the threat of administrative despotism. As noted in the Introduction, a quiet kind of democratic unfreedom pervades America today. Citizens increasingly shrug their shoulders and look to bureaucratic experts and "policy wonks" to solve any and every problem. But insofar as ecocrisis involves reconsideration of either the socially legitimated ends of the industrial growth society or public values outside the market framework, this strategy must fail. Edwin Haefele argues that environmental crisis has made all too clear the weakness of the modern bureaucratic state. "So long as the problems faced by governments could be defined as technical problems, then they could be passed along to experts and the solutions arrived at by a nonpolitical or covertly political process. But environmental problems quickly became value judgment problems. Moreover, they obviously required governmental action" (1973, 15). In chapter 5, I show further why religion is the institution best suited to deal with the ethical and political questions raised by ecocrisis, precisely because it empowers citizens to deal with the questions that elude the bureaucratic mentality. In the remainder of this section, however, I want to look at two other problems associated with the strategy of managing planet Earth.

Managerial Arrogance, Technological Optimism, and Unknown Risk

According to many critics, the strategy of managing planet Earth is one of unjustified technological optimism, human arrogance, and unknown risk: too optimistic in that the advance of knowledge cannot be predicted; arrogant in assuming that human beings have the knowledge (intelligence, financing, time, and institutional apparatus) necessary to replace the biosphere developed over billions of years of natural history; risky in that the cost of failure would be ecologically catastrophic, an irreversible blunder from which humanity, let alone the diversity of life, would never recover. Only a brief critique of this strategy is possible here, though readers might consider an array of sources, especially those critical of technology, beginning with the novel *Frankenstein* and including Heidegger's *Question Concerning Technology*. Here, however, John Firor's critique of the strategy seems particularly relevant, since his discussion is set in the context of the changing atmosphere. Firor suggests that successes in modern medicine and space science, and in controlling and remaking the face of the earth, have induced the characteristic Western mind-set that "we are skillful enough to manage everything," including global ecology (1990, 100). But, he continues, there is no evidence to support the expert's notion that humans can "replace the accumulated 'wisdom' of the interconnections between air, earth, water, and species with our own intelligence, diligence, and management skills" (103). Firor offers five reasons why we should carefully consider alternatives before proceeding down the path of expert management.

First, the path of scientific discovery and technological development (or engineering) is never as predictable as the experts claim but is always pock-marked with delays, failures, and even catastrophes. We still await the clean, cheap, and safe energy promised by the proponents of nuclear power. The near disaster of Three Mile Island and the spiraling costs of Texas's Comanche Peak nuclear power plant show that "clean, cheap, safe nuclear power" is more an entrepreneurial slogan created to curry public acceptance and support than a realistic assessment of nuclear power. Granted, nuclear power does not produce CO_2, but, like CO_2, radioactive by-products must go somewhere, and that somewhere is nature.

Second, technological fixes often fail to solve the problems they address. Consider that pesticides have led not to a bug-free environment but to generations of insects that are resistant to pesticides, to hidden and incommensurable costs to infrahuman species, and to increased human mortality from cancer. Rachel Carson exposed the underlying fallacy in *The Silent Spring*. "The 'control of nature' is a phrase conceived in arrogance, born of the Neanderthal age of biology and philosophy, when it was supposed that nature exists for the convenience of man" (1962, 261). No informed environmentalist questions the need for ecotechnologies. What is at issue is the governing mentality that directs the process of research, development, and ultimately implementation.

Cancer and other so-called diseases of civilization suggest a third weakness in the expert narrative that we can manage planet Earth. That is, technological fixes often create new problems in attempting to solve others. Cancer is one example. The so-called population explosion, termed "a raging monster upon the land" by E. O. Wilson (1992, 328), is another. The unrestrained growth of human population confirms the unintended consequences of modern technology, for no one set out to overpopulate the world. The visionaries of the modern age envisioned a world free of disease, with plentiful supplies of clean water and nutritious food for all peoples in all nations. This golden age was to be made possible through the power of technologies, such as the "green revolution." Burgeoning masses of starving human beings give lie to the vision of technologically created utopia.

A fourth flaw in the managing planet Earth scenario is that it ignores economic limits. Firor notes that "as solutions involve increasingly complex technologies and higher and higher costs, more of the world's people are left behind" (1990, 105). The Third World faces a dilemma, pursuing the economic development model exported by the West, with its adverse ecological consequences, yet suffering from an inability to finance the pollution abatement technologies needed to forestall the negative environmental outcomes of industrialism itself.

Last, and perhaps most important, the kind of world that the technological imperative or expert management path leads toward is potentially antithetical to our most basic values. Is the path we are presently embarked on, Firor asks, "leading us to the kind of world we would like to live in," or is it leading toward administrative despotism (1990, 107)? Even assuming that technology could replace the natural rhythms that govern the global climatological and hydrological cycles (in itself a major assumption), to implement such a technology would require a dramatically different social system. Murray Bookchin terms it "ecological technocracy"—one that will "require a highly disciplined system of social management that is radically incompatible with democracy and political participation by the people" (1990, 171).

Issues of Equity

The strategy of managing planet Earth avoids issues of equity between the so-called First and Third Worlds—the developed, industrialized nations and the "underdeveloped" nations. The industrial growth scenario, planned by experts and pushed into practice by politicians, assumes the inevitable solution of ecocrisis through technological fixes and economic dynamism. That story is contradicted by an array of evidence that there are no strictly technological solutions. David Victor argues that since climate change is a global problem, any approach that ignores issues of equity is simply implausible. Yet any strategy that seriously considers issues of equity confronts a host of difficult problems, since it "will have considerable effects on virtually every aspect of industrial and agricultural societies," including income transfers that may "amount to tens of billions of dollars" (Victor 1991, 452).

The relentless assault on the Amazon rainforest confirms both the influence of the Western economic development model on Third World planners and the basic failure of that model. Critics point out (1) that the development model ignores the cultural realities of the Third World (for example, that elites rather than the masses usually benefit), (2) that plans made by Western experts are usually ecological disasters in geographical and ecological context (for example, the equatorial rainforests cannot sustain either forestry or agriculture on the Western model), and (3) that development serves, more than anything else, the economic interests of the Western world, predicated upon the marketing of goods (computers, turbines, construction machinery) and experts (engineers, managers), ultimately creating permanent dependency on the West. Wilson argues convincingly that sustainable forestry is possible in the rainforest, consistent with both the need to preserve biodiversity and an extractive economy (1992, 322ff.), but this requires different conceptual strategies, such as bioeconomic analysis that considers the potential for entire ecosystems to provide habitat for species and human economic benefits on a sustainable (long-term) basis, rather than traditional economic models, which discount future value in favor of short-term growth and profit (319ff.). Wilson even emphasizes the importance of "preliterate knowledge"—that is, the concrete knowledge of rainforest aborigines—as being more important in some areas than expert knowledge (322).

E. F. Schumacher's *Small Is Beautiful* (1973) remains a revealing critique of the Western developmental model. He argues that what the economically impoverished people of the Third World require is appropriate technology, that is, technology consistent with their human and economic resources, cultural traditions, and geographical location. Rather than Western-style hydroelectric dams providing electricity for cities and water for agribusiness, the rural people of Africa need wells and filtration equipment that provide clean drinking water and supplies for irrigating local gardens. Rather than billions of dollars of financing for plants that make products for export with low-cost labor (jobs moved by international corporations from their own nations), the poor of the Third World need small loans to start locally owned businesses and craft industries. Rather than providing safe-havens for dangerous manufacturing processes (such as Union Carbide's chemical plant at Bhopal, India), moved from developed nations with stricter environmental regulations and higher costs of labor, developing nations need environmentally safe enterprises producing goods that can be consumed locally. Rather than high-technology, energy-intensive methods of mechanized manufacture that convert raw materials into finished goods for export, Third World people need systems of production that provide employment for lots of people and goods for local consumption.

The paradox of environmentalism, as I have termed it, gives further reason for skepticism about engineered solutions. Since 1970 all the Western democracies have made efforts to respond to ecocrisis. The failure to reverse the processes of destruction implies that Western culture is embedded in an ecologically pathological paradigm. There is no evidence that the scientific knowledge necessary to "manage planet Earth" exists, since neither the advance of knowledge nor the creation of technological fixes can be predicted. That the intelligentsia of Western society would want to manage their way out of ecocrisis makes perfect sense. Because that is the Western paradigm. We have been trying to manage the planet for at least three hundred years. As prophetic as it might sound, by all available evidence something more than ecotechnologies is needed if Western culture is to survive. My thesis is that this something is religious discourse, since it is the primary form of cultural conversation outside the modern story of economic growth and technological fixes.

Can Religious Discourse Make a Difference?

Religious discourse, of course, helped forge the socioeconomic paradigm that rules the modern age.[21] Francis Bacon, perhaps more than any other individual, convinced his generation that through science, humankind might recover from the Fall, might rebuild what sin had put asunder. The irony is that religion, however vital it remains for individuals, has become a dead letter at the collective level—that is, in relation to the central purpose of the modern state. For the state exists, above all else, to keep the machinery of utopian capitalism running smoothly.[22] Even state welfare programs can be interpreted, in addition to serving social justice, as serving economic ends. (See chapter 2, below, for further discussion.) Whatever the role of religion in helping shape modern society, it is largely irrelevant today to the normal operation of the state. Religion exists on the margins of the institutional apparatus and legitimating narratives that govern societal decision making. And yet, perhaps ironically, the relegation of religious discourse to the margins has left it outside the Imperium—the institutionalized, governing structure of the new industrial state.

Religious discourse remains, across the spectrum of religious belief, a second language, *a language of the heart* that speaks to purposes and gives voice to issues outside the modern materialistic vocabulary of utilitarian individualism. Bellah suggests that it is neither a foregone conclusion nor entirely clear that "Americans are prepared to consider a significant change in the way we have been living. The allure of the packaged good life is still strong, though dissatisfaction is widespread" (1985, 294). But the potential is there, confirmed by the apparent consensus on ecology, indicating that Americans are prepared to sacrifice some of the quantity of life for environmental quality, even though environmentalism has not become a political force able to empower social preferences beyond the market (Hays 1987). What is clear to many observers of the social scene is that (1) no adequate redirection of the commonweal will occur through the first language of utilitarian individualism, and that (2) certain social preferences—for example, for sustainability, for preservation of endangered species and wild places—are incapable in principle of being realized through the market. As Mark Sagoff (1988) argues, all social goods are not traded on the market; indeed, there are goods that the market thwarts. Religious discourse, and biblical language specifically, is essential to redirecting the commonweal and politically enabling social preferences the market either thwarts or endangers.

Political competency, for a democratic society locked in the grips of ecocrisis, begins with the ability to consider alternatives to the present fateful direction. Insofar as alternatives are unspeakable, incapable of being named, then a democratic society cannot adequately discuss solutions. Manfred Stanley argues that *a politics of competence* "begins with the question of whether we can any longer afford to allow agents of powerful institutions to do anything they want with language, the most significant of all human symbolic phenomena" (1978, 251). Regrettably, powerful politicians, philosophically aligned with the corporate sector and the community of experts, have systematically engaged in a politics of deceit, managing through rhetoric to turn the tables on environmentalism during the 1980s and early 1990s.[23] The problem is not the cupidity of politicians per se: it is the modern paradigm. That paradigm articulates itself through utilitarian individualism—the language of *Homo oeconomicus:* our first language, institutionalized through our political economy. As many, many scholars agree (Bellah, Johnston, Galbraith, Heilbroner), thinking about our society and its direction is narrowly focused on the economy. This is not surprising, since the new industrial state, as Kenneth Galbraith calls it, or the era of controlled capitalism, as others term it, dictates governmental social policy and thereby influences virtually everything else, including our character. When it comes to ecology, we all have feet of clay.

We also have a second language, rooted in our biblical and republican traditions, that can challenge the first language of utilitarian individualism. Biblical and republican values, in other words, or communities of memory, can countervail our first language. Crucially, for most Americans religious discourse is the only language that remains outside the modern paradigm. It is therefore essential to reaching conceptions of the public good beyond narrow economic definitions restricted to monetary measures. Although they may seem impervious to change, our political economy and the associated complex of institutions (stock markets, Congress, corporations) and laws can be changed. Social movements have time and again caused fundamental transformations in public policy, the civil rights and Vietnam antiwar movements being recent and vivid examples (occurring within the lifetimes of many Americans of voting age). In short, transforming American culture necessarily involves its *biblical and republican traditions.*

Barring the cultural quantum leap envisioned by New Age apologists and gurus, a revolutionary change in consciousness that transforms American culture overnight, there are no realistic alternatives.[24] The road from here to there—from a cultural paradigm that endangers itself by destroying nature to a sustainable culture that is in dynamic equilibrium—begins with available resources. Of these, none presents a more potent possibility than religion, since no solution for ecocrisis within a practical time frame can be conceived outside the presently existing cultural framework. Further, argue close observers of the political process such as Robert Paehlke (1988), there is little if any likelihood of a green party majority anytime soon. Americans lack the time to reinvent the political process or the two-party system. The problem is to find the means of expression that will enable a politics of competence. Environmentalism, therefore, insofar as it is to become effective, must achieve its agenda through existing cultural resources. The question is whether religious discourse can produce the solidarity necessary to redirect utopian capitalism.

The idea that religious discourse could do so seems nonsensical at first. Religion today appears to be entirely a matter of private values and personal affairs. Further, the root values of Judeo-Christianity seem to many to have been co-opted by the gospel of greed—the idea that financial success is a sign of divine favor. And the separation between religion and politics seems sacrosanct, a line that no politician—even one with an environmental agenda—dare cross. Political candidates, particularly presidential candidates (the president and vice president are our only nationally elected officeholders), have to pass the litmus test of neutrality.[25] Americans seem to agree that the state is one thing and religion another. By tradition and law, church and state are separate. Any claim that religion might influence policy, albeit policy concerned with common and therefore public goods, seems a heresy. The state, insofar as it concerns religion at all, guarantees freedom of worship. Otherwise the state simply attends to its own functions, keeping the machinery of capitalism running, providing for national defense, building the infrastructure of public works, and so on. A closer look, however, reveals that these objections are not insurmountable obstacles.

The idea that religion might give environmentalism political potency should not be dismissed out of hand. Religion has played a crucial role in national affairs time after time. Martin Marty points out, for example, that "the black civil rights movement was a development of the stories of Sinai and the Promised Land, of American slavery and liberation, of suffering and triumph, of evil surroundings and potential virtue in the activities of the black community and its allies" (1989b, 13). Granted, there is no U.S. state religion, and that reality underlies the legal separation of church and state. But voter preference can be and often is shaped by religion, and through the exercise of the vote religion influences the public agenda. The Great Code undergirds a second language that enables the possibility of identifying issues and ultimately conceptions of the public good. And these conceptions might lead toward a socially just and ecologically sustainable society. There is arguably a common interest in caring for creation across the entire spectrum of religion in America. If the faithful express this commitment, environmentalism might achieve the moral authority it lacks.

Beyond the ability of religion to affect the political process lies a deeper reality. Channels laid down by the Great Code, however obscure or hidden from view they may be, still exist; and we flow in them. In the West, Judeo-Christianity has been and remains the most important influence on human values. To use the language of sociology, religion is the shaper and weaver of the sacred canopy, the framework of ends that gives meaning to human existence and motivates people to act.

Even granted, however, that religious discourse remains free of utilitarian individualism, many obstacles within the church (organized religion) make any claim for its role in a time of ecocrisis suspect. Many, including some within the frame of Judeo-Christianity, argue that the patriarchal nature of religion precludes meaningful change, since God is placed over man, and man over woman, children, and nature. The emphasis of some Protestants on salvation seems to confirm the charge of environmentalists like Lynn White and John Livingston that Judeo-Christianity is too otherworldly to make a difference. Even the pope's call for Catholics to treat nature with care and respect seems hollow to some environmentalists, given the Catholic position on birth control.[26] And some believers think that environmentalism is yet another outbreak of paganism, part of the devil's work, a ruse to snare unsuspecting souls to care for nature rather than God.

A future generation of historians will perhaps be able to write the history of this decade, recounting the role that Judeo-Christianity played or did not play in ameliorating our ecocrisis. The question is empirical: *Will the faithful rise to the occasion?* This book is not a prophesy that believers will be successful but a conjecture that they not only might but have already begun to develop religiously inspired environmental ethics. I remind the skeptic that pessimism is a self-fulfilling prophecy and that environmentalists have too often argued against a caricature of religion. Even while claiming that religion is a necessary condition for the solution of ecocrisis, any cogent argument must almost simultaneously recognize that technology, science, conservationists, and other forces serve essential functions. Environmentally relevant education is also necessary; its importance should not be underestimated. Religion alone is not the solution, but if the role of religion is recognized and acted upon, there is a chance that we will escape the paradox of environmentalism.

2

Religion and the Politics of Environmentalism

Religion has been at the center of our major political crises, which are always moral crises—the supporting and opposing of wars, of slavery, of corporate power, of civil rights, of sexual codes, of "the West," of American separatism and claims to empire. If we neglect the religious element in all those struggles, we cannot understand our own corporate past; we cannot even talk meaningfully to each other about things that will affect us all.—Garry Wills, *Under God: Religion and American Politics*

Surveys of attitudes on environmental issues are difficult to interpret, partly because the results across surveys appear equivocal and partly because voter preferences seem to be volatile. Presidential campaigns support this conjecture. During the 1992 primary season, unlike the situation in 1988, economic issues overshadowed all others. A CNN/*USA Today* poll in late 1991 ranked environmental issues eleventh out of fifteen items of national concern (Loftis 1992, 19). A poll taken in early 1992 indicated that 51 percent of the voters thought environmentalists went too far in making demands on industry and government, while 42 percent thought they did not go far enough (C. Alexander 1992, 51). Charles Alexander interprets these data as an economic backlash against environmentalism, led by political conservatives and the so-called wise-use movement—a variation on the progressive conservation movement. He argues that the only way environmentalism can regain the initiative is "to show more flexibility and demonstrate that conservation is not incompatible with economic growth" (1992, 52). However, the same Yankelovich-Clancy-Schulman poll showed that 50 percent of the electorate thought that funding environmental cleanup should be continued, regardless of such other issues as the economy.

The paradox goes beyond these recent polls, however. William Ruckelshaus, chief administrator for the Environmental Protection Agency from 1970 to 1973 and again during 1983–84, notes that even though a majority of voters had a pro-environmental stance during the 1980s, public policy lagged far behind.[1] While polls indicate that almost 80 percent of voters believe that "protecting the environment is so important that requirements and standards cannot be too high, and [that] continuing environmental improvements must be made regardless of cost," legislative action has not followed public opinion (Ruckelshaus 1989, 169). The problem is that pro-environmental legislation often has adverse economic effects on special interests. As Ruckelshaus points out, the so-called politics of interest, as distinct from the politics of the common good, triumphs, for the obvious reason that a "much injured minority proves to be a more formidable lobbyist than the slightly benefitted majority" (169).

Although long a matter of interest to scholars, the political influence of special interests has typically been ignored by the American public, which prefers to believe that the public interest somehow always wins out. The paradox of environmentalism suggests otherwise. During the presidential campaign of 1992 the problem of special interest politics came dramatically to the front with the third-party candidacy of Ross Perot. Although Perot's primary concerns were economic and budgetary rather than environmental, his vivid descriptions of forty thousand lobbyists wearing alligator shoes and thousand-dollar suits sensitized the electorate to the force of special interests in determining the public agenda. Special-interest politics is, in fact, the rule rather than the exception. Most Americans, as Perot tried to make obvious, do not realize this.

The problem of special-interest politics is exacerbated by voters' tendency to reject ecologically prudent policies that affect their way of life. For example, even though automotive emissions are a leading source of greenhouse gases, there is no indication that Americans at large perceive mass transit as either desirable or even feasible. Neither does there seem to be a public clamor for the development of alternative technologies, such as hydrogen-fueled automobiles, that would help reduce CO_2 emissions. In part this is because such alternatives would, at least initially, raise the cost of transportation. When consumers avoid paying the premium for devices, such as automobile air bags, that are in their own selfish interest, why should anyone expect them to pay for automotive technology that primarily benefits future generations of human beings and other life-forms? Ecological altruism, defined as the coincidence of self-interest with other-interest, does not seem likely, if for no other reason than economic self-interest trumps the willingness of individuals to pay for their own benevolent interests in future generations or other species.

The politics of interest, then, seems to cut on both sides of the market society, since producers and consumers alike appear zealously ready to defend their selfish economic interests over any altruistic interest in public goods like environmental preservation. As noted, the vast majority of the citizens of Western industrial democracies are embedded in the Dominant Social Matrix (DSM; see table 1). The DSM presents a challenge to my thesis, since Americans, regardless of their faith commitments, apparently prefer to live in ways that require massive consumption of natural resources and generate huge outputs of waste (pollution).

Demographic and economic facts appear to confirm this judgment: Americans constitute less than 5 percent of the world's population but consume more than 30 percent of its resources. And given the paradox of environmentalism, it appears that organized religion is incapable of transforming the DSM. Although 90 percent of us have a religious affiliation, most of us apparently think of our religion as a private matter with little or no relevance to public issues. The pervading American interest in possessing things (materialism) is taken for granted, even to the extent that some believers see wealth as a sign of God's favor. Douglas Bowman argues that the church itself appears to be embedded in the DSM, since it tends "to avoid the urgency of the environmental crisis. . . . We push the awful prospect of environmental catastrophe to the back of our minds and get on with other tasks and other thoughts" (1990, ix–x).

Table 1 The Dominant Social Matrix (DSM)

- Nature has instrumental (anthropocentric) value only; biocentric values, such as the preservation of endangered species, are meaningless.
- Short-term economic interests override long-term issues like intergenerational equity; future generations of human beings will be able to fend for themselves.
- If environmental risks caused by habitat modification, consumption of resources, and the emission of pollution are economically beneficial (as measured monetarily), then they are acceptable.
- Environmental risk poses no limits to growth, just problems that require engineered solutions (for example, restoration of habitat, resource substitution, and pollution control technologies).
- The strategy of managing planet Earth is feasible; through biotechnology and other sciences, humankind will ultimately be able to control all biophysical processes on the planet.
- The politics of interest is sufficient to guarantee that the best available technology to restore habitat, devise resource substitutions, and control pollution will be employed.

But there is reason to think that the hegemony of the DSM is being challenged by a New Social Matrix (NSM; table 2).[2] Clearly, given the twin realities of the paradox of environmentalism and the politics of interest, proponents of the NSM are a minority. History reminds us, however, that the DSM itself was once a newcomer, a challenger to the world-in-force. As Bowman argues, "If we think of the world as an impersonal resource—there for the taking—we will continue to live like ingrates and exploit the biosphere like parasites." But religion and the church can play a role in leading the way to the New Social Matrix. "If we view the earth as a gift," Bowman continues, "a fragile child of God, a nurturing fellow creature, a new 'Christ' wounded by humans and bleeding on a cross, we will discover a new reverence, and our lives will reflect that altered habit of mind and spirit" (1990, x). Bowman's position, obviously, is more restrictive than mine because he is addressing Christians exclusively, but his point is apropos of my thesis.

Table 2 The New Social Matrix (NSM)

- Nature has intrinsic value (value in its own right apart from human interests) as well as instrumental value; a healthy economy cannot be sustained by a sick environment.
- Long-term issues, such as intergenerational equity, are at least as important as short-term economic interests; sooner or later someone pays the costs associated with short-term greed, and that someone is our children and the infrahuman species adversely affected by actions motivated solely by economic self-interest.
- Economic activity always entails risks, but risk that entails either unpredicted or irreversible ecological consequences is not acceptable no matter how profitable; when in doubt—and especially when human action affects fragile ecosystems, endangered species, or has global implications—act conservatively.
- There are biophysical limits to growth that no human technology can overcome in the long-term, though these limits can be exceeded in the short-term; accepting limits to growth does not mean that human beings live in degraded conditions of poverty. A life of relative plenty is possible even in delicate ecosystems.
- Creating a sustainable society is a feasible alternative to the modern project that attempts to manage planet Earth; the modern project is in fact driving the earth toward ecocatastrophe. Hubris sustains the illusion that humankind can control the biophysical processes that govern life on earth.
- A citizen democracy, attentive to local geography and environmental issues as well as to global issues, is required to build a sustainable society that is also consistent with democratic life.

Environmentalism as a Social Movement

Whatever the future might bring, there can be little doubt that the politics of special interest now dominates the political process. Claims about voter sovereignty are one thing; the reality of political decision making is another. William Greider (1992) argues that the governing classes of professional politicians and special-interest groups have almost totally assumed control of the political process. The average citizen has been systematically disenfranchised under the DSM—pushed to the periphery of societal decision making. The point is made succinctly by R. J. Johnston: "While elections produce governments, it is interest groups that influence them" (1989, 167). There is no mystery about this. It is primarily because the politics of interest is entirely consistent with the prevailing public philosophy, utilitarian individualism.[3] Private interest has had and will continue to have great political influence—and legally so. Grant McConnell (1966), among many others, has shown that democracy works very well to serve private interests. The common good is more often than not an afterthought, the rhetorical veneer covering "public policy" that economically subsidizes and politically advances private interest. History confirms the repeated co-option of the public interest by private power, especially (although not exclusively) in regard to environmental issues.

Majority rule is, like voter sovereignty, a useful fiction, but on specific environmental issues, such as water policy, forestry, or pollution control, majorities rarely rule. (For a case study, see Donald Worster, *Rivers of Empire*.) G. William Domhoff, in his study of the political process, pessimistically concludes that there is an effective ruling aristocracy in America "based upon the national corporate economy and the institutions that economy nourishes" (1967, 156). Johnston suggests the dominant ideology of utopian capitalism reduces elections to a contest over which party is most "able to operate the large and complex state apparatus and sustain it in its basic functions of securing social consensus, promoting capitalist accumulation, and legitimating the mode of production" (1989, 158). Contrary to appearances, there is little if any real debate over the public interest in environmental issues. As Robert Dahl points out in *Democracy and Its Critics*, much skepticism about democratic life stems from our political economy's institutionalization of private, narrowly economic interests. Special interests overdetermine public policy.

Pessimism about the potential for democratic process to surmount the politics of interest is a self-fulfilling prophecy. *Religious discourse, expressing itself in the democratic forum, offers the possibility of overcoming special interest politics—especially those which are narrowly economic—on environmental issues.* Clearly, the diversity of religious belief, the separation of church and state, the privatization of religion, and the materialistic tenor of American society makes the actualization of this possibility less than a foregone conclusion. My argument is conjectural, although there is evidence that a religiously inspired environmental movement is in the first, tentative stages. If a consensus on sustainability can be developed, then modifying the market through policy that reflects majority opinion is inevitable. As advocates of both the DSM and the NSM agree, environmentalism unquestionably has the potential to lead to fundamental social change. Anna Bramwell, who supports the DSM, believes that a "green consensus" on the NSM would destroy the basic character of Western civilization. Supporters of the NSM are more positive about the future. Ruckelshaus, for example, argues that sustainability entails redefining "our concepts of political and economic feasibility" as well as modifying existing and devising new economic and political institutions. Such concepts and institutions are, he continues, "simply human constructs; they were different in the past, and they will surely change in the future" (1989, 174). Will Wright (1992) goes even further, arguing that sustainability as such is a necessary characteristic for any viable social narrative. On his view either the transformation from the DSM to the NSM occurs or society collapses.

Surveys indicate that Americans in general say they are willing to pay more to protect the environment and that they will do so even if economic growth must be redirected or more closely regulated.[4] But as Samuel P. Hays (1992b) points out, voter support for pro-environmental candidates varies regionally, largely according to the differing nature of local economies. Texas, for example, was Reagan country in the 1980s, as is predictable from the extractive and resource-intensive nature of its economy. Voter support also varies through time; ecocatastrophes (like the wreck of the *Exxon Valdez* in Prince William Sound) lead to a surge of media attention and public interest (including increased membership in mainline conservation organizations) that almost immediately lags as other issues, like the economy, recapture public attention. Lance deHaven-Smith argues that there is no evidence that Americans generally are committed to any environmental ethic or philosophy. People who "vote green" do so largely because of local issues that adversely affect the quality of their lives. The Reagan anti-environmental revolution (Hays 1987, 491), then, reflects the fact that any collective interest in environmental quality is often offset by the economic interests of specific groups. More generally, although voters are willing to pay for local improvements in environmental quality, they do not seem willing to pay for the costs of creating a sustainable society.

There is a basic problem, however, in equating environmentalism with a "willingness to pay for protection"—*the standard economic interpretation of environmentalism*. This economic model of environmentalism can be construed as more symptomatic of ecocrisis than a cure for it. Simply stated, a cost-benefit approach to environmentalism assumes that public debate over and solutions for ecocrisis can be dealt with adequately through a quantitative (economic) approach. Such a strategy is inadequate, because a quantitative approach glosses over the qualitative issues, that is, *the ethical and normative issues*, that ecocrisis raises (Wilson 1992; Firor 1989). Utilitarian individualism is the prevailing public philosophy, yet a society that decides its future on those terms commits "the utilitarian fallacy"—that is, it succumbs to the belief that the only real values are those that can be quantified through the market (so-called shadow pricing).[5] As Eugene Hargrove argues, modern economics is a direct outgrowth of utilitarian individualism, and it promotes the ideal that "rational egoism, or selfishness" is not only "an objectively valid approach to ethical action" but "the only rational approach to human action in general" (1989, 209). Cobb and Daly assert that market economics and cost-benefit analysis may be rational for individuals, but for societies they are irrational if for no other reason since, in contrast to individuals, communities are quasi-immortal. "Social decisions therefore should be discounted at zero insofar as mortality is concerned" (1989, 152). The willingness or unwillingness of the present generation to pay for preservation reflects a discounting of the value of future life. But a culture goes on: to discount the future value of old-growth forests, clean air, energy conservation, and so on across the panoply of environmental issues is perhaps economically expedient (given the dominant form of economic theory) for the present generation but it is potentially disastrous in the long run.

In contrast to the standard economic interpretation of environmentalism, some analysts believe that pro-environmental public opinion manifests a growing coalescence of interest around a *public point of view that utopian capitalism is undercutting nature's economy*. Sagoff believes that environmentalism cannot be adequately conceived in terms of the market and consumer willingness to pay for clean air, wilderness refuges, or the protection of endangered species. Problems like these, he argues, "are primarily moral, aesthetic, cultural, and political" (1988, 6). Some welfare economists (although they are a minority) concur, arguing that simple trade-off models ("the willingness to pay") and cost-benefit analysis of public policy issues ignore critical non-market effects, such as the quality of life, the value of nonhuman ecology, and the worth of beauty. That cost-benefit analysis discounts the future raises serious and unresolved ethical questions about its practice. Clive Splash argues that "discounting the future at almost any positive rate creates insignificant present values for even catastrophic losses in the distant future" (1993, 119).

Election outcomes in the early 1990s, such as the defeat of the so-called Big Green initiative in California (1990), and the actions (or inactions) of the federal government over the past decade imply that the path from any "public point of view" on the environment to the election of pro-environmental candidates and support of pro-environmental policy initiatives is not smooth, even for a single nation, let alone the entire human community.[6] Granted, citizens *appear* prepared to support pro-environmental legislators and public policies. A few green senators and representatives have been elected. Yet in spite of positive achievements, such as the National Environmental Policy Act and clean air legislation, the relentless growth of the most salient indexes of environmental degradation—whether we choose to look at environmentalism through the market and willingness to pay or whether we understand environmentalism as portending fundamental societal transformation—imply that whatever the promise, the era of green politics has not arrived. Presidential candidates who have advocated serious environmental policy changes, among them Stewart Udall in 1976 and Al Gore in 1990, have not been well received. The situation may change, but we live in a time of flag-draped politicians who offer us "environmental sound bites."

Robert Paehlke argues that environmentalism might be the catalyst that rejuvenates the Western world's liberal democracies and leads them to a new world order of peace, economic equity, and sustainability (1988, 282). In part this is because environmentalism necessarily involves a discourse on values, the articulation of collective preferences, the discussion of a public point of view. Samuel Hays ties the rise of environmentalism as a post–World War II political force to changing public values, especially concern over environmental quality (1987, 1992a, 1992b). Yet the paradox of environmentalism and the anti-environmental policies of the Reagan and Bush administrations demonstrate that environmentalism has not arrived as a moral force adequate to the political task of moving America toward sustainability.

Michael McCann's critique of environmentalism in *Taking Reform Seriously* (1986) perhaps explains why environmentalism has failed to move society toward sustainability—and is more incisive than the usual run of anti-environmental critiques. The paradox of environmentalism, argues McCann, exists because environmentalism's goals are "limited in relevance for purposive moral direction," being largely a litany of no-growth bromides vitiated by a deep "ambivalence about the fruits of human artifice" (252). In short, environmentalism is misanthropic, lacking authoritative moral purpose. What public interest environmentalism needs, according to McCann, is a "teleological conception of human nature" that celebrates "creative social power" (253). Today's environmental leaders and groups are infected with "a crisis of confidence about even the possibility for a movement of collective purpose, vision, and grand design which might transcend their piecemeal, legalistic approach. Obsessed with preventing the worst of possibilities under existing arrangements rather than [with] restructuring society itself, the liberal ethic of preservation has ended up being largely conservative in character" (254–55).

If environmentalism is to be anything more than quick technological fixes for ecological wounds, the treatment of symptoms rather than causes, and if responses to environmental problems that scientists like E. O. Wilson and John Firor recommend are to be acted on, then Americans need to discuss the moral ends that environmentalism involves and, more generally, to engage in an open-ended conversation about the good society. As I read McCann, there is no escaping the paradox of environmentalism until such a cultural conversation ensues. So construed, environmentalism necessarily involves the attempt to articulate our choices as citizens, to express preferences for public goods that go beyond narrow economic concerns or our interests as consumers. Religious discourse converges with our republican tradition—that is, the belief that citizens are motivated by something more than economic self-interest. Sagoff's reading of the issue is unsurpassed: "Our environmental goals—cleaner air and water, the preservation of wilderness and wildlife, and the like—are not to be construed, then, simply as personal wants or preferences; they are not interests to be 'priced' by market or by cost-benefit analysis, but are views or beliefs that may find their way, as public values, into legislation" (1988, 28).[7]

The Debate over Green Politics and Democratic Process

Religion is vital to environmentalism conceived as a social movement that might transform our society's relations to nature, much as the civil rights movement transformed race relations. The civil rights movement gained influence through its claim to moral legitimacy on the basis of Judeo-Christian principles, so that its agenda ultimately (if imperfectly) became a social reality. So, too, it must be with environmentalism. Obviously, more than religion will be involved in any truly adaptive response to ecocrisis, including environmental law, technology, and science. But it is the peculiar ability of religion to promote solidarity. As Bruce Lincoln puts it, "Religion powerfully promotes social cohesion and sentiments of common belonging" (1989, 89). Religion generally promotes a greater concern for the common good, for the Creation and its ongoing processes, than any other form of discourse or association. Of course, there are exceptions. Religion sometimes promotes violence and conflict, as in Northern Ireland and the Middle East. And in Europe, with perhaps the exception of Italy, religion appears to be without significant political influence. But in the United States, unlike modern European democracies, where political parties have led the way to societal transformation, religiously inspired social movements have had considerable political influence.

Religious discourse is arguably more important than any other kind in clarifying values and preferences and, ultimately, in shaping voter choice. Obviously, environmentalism is informed by secular groups, such as the Sierra Club and the Wilderness Society. Just as clearly, religion has not to this point played a significant role in environmentalism. But that is beginning to change. Bellah suggests that American social movements develop "in the relatively unstructured public spaces in which opinion is formed but have often drawn leadership and support from churches and other established groups" (1985, 212). In our society the social movement operates, much like the church, in a middle zone between individualism or private power, expressed through the market and corporation, and public power, exercised through the institutions of government at national and other levels.

A useful comparison can be made between my thesis that religious discourse is essential to environmentalism and Paehlke's analyses of environmentalism and politics. He characterizes democratic society in terms of five so-called first-order political concerns. The first four are traditional: national security, economic prosperity, social justice, and democracy. The last, environmental protection, is the newest. Arguing that it is dangerous to grant ascendancy to any one first-order value, Paehlke nevertheless concludes that "in examining the relationships between environmental protection and each of the four other values, democracy is the guiding value, *both as an end and as a means.* Environmental protection, in particular, will be most effectively achieved through the maintenance of, indeed the continuing enhancement of, democratic practice" (1990, 349, my emphasis).

All this I accept, and indeed I would place (if possible) a greater emphasis on democratic process, an emphasis much like Robert Dahl's in *Democracy and Its Critics.* Dahl suggests that we think of the common good not as some predetermined end toward which we move by policy-making but as an issue-specific process of discussion and negotiation that culminates in the election of officials and passage of public policies to settle the matter. From this vantage point, democratic process has survival value, since a healthy community is one that moves effectively to resolve economic, social, international, and ecological problems. Because ecocrisis threatens the continued existence of a society almost as surely as invasion by an army, private interests that defeat the working of democratic process toward the ecological common good are maladaptive. *A society that cannot achieve solidarity on sustainability is self-defeating: all its primary values, including its economic ones, are ultimately thwarted.*[8] Even primary values, such as economic growth, are not self-contained absolutes, since they ultimately depend on ecosystemic integrity. As Holmes Rolston III argues, the bottom line ought not to be black unless it is also green, since "there is no such thing as a healthy economy built on a sick environment" (1986, 172).

The problem immediately encountered by the process view of democratic life and the good society is that "a political community large enough for its political life to be vital to its citizens is likely also to be so large as to include within it a variety of associations and—precisely as Rousseau feared—its citizens will hold conflicting views about what constitutes the common good and the policies that will best achieve it" (Dahl 1989, 301). *Associational pluralism* is one way to describe the problem. To say that the politics of interests overrides considerations of the common good is another. One primary value, the economy, seems to override all others. From an environmental perspective, the question is how can a pluralistic society with an enormous variety of economic associations (primarily for-profit corporations and consumers) achieve consensus on the ecological common good? If the advocates of the DSM are correct, there is nothing to worry about. The market and special-interest politics will see us through. If the advocates of the NSM are correct, associational pluralism, or politics as usual, promises to undercut the sustainability of the American experiment in democratic life.

Where I go beyond Dahl, McCann, Paehlke, and Sagoff is in emphasizing the importance of the biblical tradition to environmentalism. Whatever other modes of legitimating narrative exist, the West was and is, as Northrop Frye insists, a biblical culture. The Great Code, as Harold Bloom (1989) might put it, overdetermines us. Which is to say, apropos of the political process, that religious discourse is the primary influence on the Western worldview, such as our beliefs that time is meaningful, that it is going somewhere, and that there is a difference between right and wrong, good and evil. The Great Code, in other words, conditions the basic narrative that guides the community in its process of discussion.[9] No specific meaning can be attached to the common good apart from the community and its story sources.

Unlike Paehlke, who sees environmentalism as heralding a revival of democratic politics, Anna Bramwell (1989) associates it with green fascism, or the destruction of democratic process. Her reading of environmentalism is largely in opposition to Paehlke's (although her focus is on green European political parties rather than the environmental movement in North America). Like Paehlke, she dramatizes the revolutionary portent of environmentalism—which she calls *ecologism*, the combination of ecological science with a political agenda—to restructure growth societies into sustainable societies. But unlike Paehlke, who argues that environmental protection is a first-order value, Bramwell believes that the political implications of ecologism are destructive. She argues that ecologism, if politically efficacious so that both elections and public policy move in a green direction, is tantamount to the destruction of Western civilization. The environmental movement represents something "deathly," "a return to primitivism," "an utter rejection of all that is, and for at least three millennia all that was" (248).

Bramwell is not alone in her concern. Environmentalism has been criticized by many proponents of the Dominant Social Matrix, who believe that any challenge to the Enlightenment narrative (utilitarian individualism) is suicidal. Some critics, such as William Tucker (1982), argue that environmentalism is no more than a narrow class interest, a campaign by the middle and wealthy classes to preserve wilderness as their playground while the economic needs—which require development of natural resources—of working-class and disadvantaged people go unmet. Others, such as Julian Simon and Herman Kahn (1984), advance the thesis that the dynamic interaction of the market with technology produces solutions to environmental problems. Still others, among them Dixy Lee Ray (former chair of the Atomic Energy Commission) and Lou Guzzo, argue that the environmental movement is so out of touch with reality that it "seeks development of a society totally devoid of industry and technology" (1991, 29). Martin Lewis (1992) argues that radical environmentalism (so-called deep ecology) portends disaster, since the only solution for ecocrisis is to patch up capitalism. Even members of the American deep ecology movement, such as Michael Zimmerman, have expressed reservations about environmentalism. Zimmerman (1993) fears that, whatever the dangers posed to the earth by industrialized economies, environmentalism itself poses a threat to human freedom. For him the Enlightenment definition of liberty—freedom from government interference—is threatened by environmentalists who seek to place limits on economic growth and development. Green politics portends what he calls "ecofascism."

These criticisms ignore the workings of democratic process (as described, for example, by Dahl, Hays, and Spretnak), the traditions of democratic life, and the writings of most environmental theorists. Granted, political uncertainty is part of life and fascism a possibility. But the term *fascism* conventionally connotes aggressive nationalism, militarism, and racism—words that are somewhat descriptive of the less-than-ideal features of twentieth-century America. Further, fascist societies constrain the range of discourse through such tactics as physical coercion and censorship of the media. Rather than constraining discourse, environmentalists characteristically attempt to open or enlarge the agenda (whatever, *pace* McCann, the insufficiencies of their proposals). Writing on green politics by insiders emphasizes democratic values and the decentralization of political power—in short, a movement from the present paradigm of controlled capitalism, which ironically enjoys a "fascist like" control over the political process through the power to fund special interests, toward *citizen democracy*. Crucially, green politics also portends favorable changes for economically disadvantaged Americans, who disproportionately suffer the adverse consequences of pollution in their daily lives. Because of these health issues, the poor have a more immediate interest in creating a sustainable society than any other class; the middle class tends to move away from degraded urban habitats.

Bramwell claims that environmentalists "with a moral critique" (whom she labels with the pejorative appellation "radicals"), that is, a value-laden discourse that rejects the standard economic interpretation of environmentalism as willingness to pay, "need to explain the practical effects of their policies more stringently than parties who believe in continuing the old system, or in muddling through somehow, in being pragmatic, unideological, 'natural' in short. As soon as ecological movements become political parties they have to turn against their own values, and the more 'apolitical' in party terms they aim to be, the more ideological they become" (1989, 241). She worries that the writings of environmentalists that "concern the reconstruction of the public sphere are worryingly imprecise. Imprecision of word can arise from vagueness of thought. It can also arise from a desire to manipulate language" (247). Her fear is that *economic ecologism* "is intended to create a world where symbols produce an instant response" (247). While it cannot be fairly said that Bramwell ignores ecocrisis entirely, her position is undemocratic, since she overlooks the stubborn reality that environmentalism is more than anything else a debate over the common good. Bramwell must surely admit that a society which is not sustainable is self-defeating and therefore not a good society. By categorizing environmentalists as radicals, she aligns herself philosophically with the

DSM and politically with economic special interests. "Muddling through," as Bramwell terms it, means that the economic interests of the present generation of consumers and producers either supersede or take precedence over other primary interests, such as democracy and sustainability.

Although Bramwell is off the mark in her assessment of the threat "environmentalism" poses to Western civilization, she helps us focus on the importance of discourse—the democratic discussion of an environmental agenda. Clearly, she wishes to restrict the range of discourse by favoring the DSM over the NSM. But environmentalism, if anything, promises to expand the public conversation to include a larger domain of value considerations than those given voice by utilitarian individualism. Bramwell's transcendental signified is the modern paradigm itself, as epitomized by Adam Smith. Her view of human beings, stripped of its rhetorical veneer, is that they are ultimately nothing more than greedy little pigs. Traditional economic theory is thereby a privileged form of discourse: a cognitive absolute used to judge all other positions. Like Adam Smith, Bramwell cannot envision a political economy as pursuing anything other than quantitative growth. The Gross National Product is her God.[10]

The movement toward sustainability (which could be measured by reductions in such indicators as atmospheric CO_2, population growth, and so on) is thwarted by utilitarian individualism generally and more specifically by the ability of special-interest groups to maintain their privileges. Most people, as Ruckelshaus (1989) suggests, know this. Yet green politics appears to be subjective, involving soft and ambiguous goals in contrast to the quantifiable ends of the political and economic establishments. Economic interests are after all "objective," especially when compared to environmental values.[11] Economic statistics, such as the rate of growth in the GNP or per capita income, make good sound bites that Americans are conditioned to accept as intrinsically meaningful, hard facts about the world. In contrast, talk about the long-term implications of climate change or the extinction of species appears qualitative, subjective, and idealistic. Accordingly, experts attempt to reduce ecological issues to economics. And this undercuts any public discussion of an environmental agenda that addresses the moral issues involved in creating a sustainable society.

Michael Nieswiadomy claims, for example, that environmental problems, correctly understood, "are fundamentally economic problems, because they result from the economic decisions of individuals" (1992, 120). The public needs, he continues, "to be wary of proposals offering the socially acceptable options. . . . Good intentions alone will not solve our environmental problems. Environmental economics offers a methodology to prudently manage our natural resources and environmental problems" (130). The problem with the standard economic interpretation of environmentalism, as Sagoff points out, is that from its perspective "there is really only one problem: the scarcity of resources. Environmental problems exist, then, only if environmental resources could be used more equitably or efficiently so that more people could have more of the things for which they are willing to pay. . . . The only values that count or that can be counted, on this view, are those that a market, actual or hypothetical, can price" (1988, 26–27).

If environmentalism is to be effective in the United States, then solidarity on a social agenda is required. More accurately, solidarity is a necessary condition for creating a politically efficacious response to ecocrisis. Notwithstanding the arguments by intellectual elites within the environmental movement, who herald the coming reality of a paradigm shift that will suddenly transform society, the road from here to there, in my opinion and the opinion of many others, can be reached only through existing means. Thus, as Milbrath argues, "Even though [opinion polls indicate that] the public favors environmental protection over economic growth, large differences between the [environmental] rearguard and vanguard within countries signals that this will be an issue of strong political contention for many years to come" (1989, 124). Green politics, Milbrath argues, is an ideological struggle between advocates of the DSM and the NSM for the "undecided middle" of voters (134). What Milbrath does not emphasize is that debate itself, a public conversation on environmental issues expanding discourse beyond economic confines to include ethics, is the key to achieving solidarity.

Both environmentalists and scholars in environmental studies tend to believe that simply discussing the underlying fallacies and limitations of the industrial growth paradigm will be sufficient to create change. This approach ignores the long-standing equation of economic growth, as defined in terms of neoclassical theory, with social progress. Economic growth is not only socially legitimated but also institutionalized in American culture as the standard of progress. There are definite correlations between quantitative economics and human well-being. Poverty destroys lives, spiritually, physically, and socially—on the Horn of Africa and in the ghettos of America. Just as obviously, there is no law of nature (as economics pretends) that all economic growth improves life. As mentioned, a number of issues are raised by exclusively monetary measures of growth, such as who benefits and who pays, hidden environmental costs, and costs passed on to future generations. When these questions are not asked, the GNP and rate of economic growth become secular absolutes beyond criticism.

My argument is that, when it comes to governance, economic policy is a secular religion, and the GNP and rate of economic growth are our holiest of holies. Almost every American has been socialized to believe in them as virtual absolutes. Indeed, even from the beginning of our history, economic progress was a moral ideal. As Bellah notes, "The United States was planned for progress. It was a commonplace among the country's eighteenth-century founders that economic life was not simply a neutral fact of nature, that political economy was a branch of political ethics and its practice an exercise in public morality" (1991, 67). This belief, equating economic growth with moral progress, came from John Locke (and was refined by Adam Smith).[12] But Locke's exaltation of private property and individual rights was grounded in a Judeo-Christian narrative tradition, "inseparable," as Bellah puts it, "from his theology and from his stern Calvinist sense of obligation" (67). The economic system Locke envisioned, then, existed within a system of religious constraints. But the system that Locke envisioned, in which religion framed political economy, was eroding by the nineteenth century. Today religion and economics are divorced.

It is precisely at this juncture that environmentalism transects the biblical tradition. Of course, to the secular humanist such a contention is nonsense. God is, after all, dead, and religion is more problem than solution. What the secular "green advocate" ignores is the cultural context that enframes environmentalist discourse. If we think of ourselves as storytelling culture-dwellers, it is easy to see how the Great Code influences us. Culture can be described in various ways, but clearly it is a web of talk, a continuous conversation about how to live. Religious discourse is an essential voice in that conversation. Biblical language has been a vital part of our nation's public debate over the structure and texture of the good society. There is no reason that it cannot play a role in determining an environmental agenda.

Those who reject the standard economic interpretation of environmentalism, such as Sagoff, believe that the refusal to think of environmentalism in terms that go beyond the market is deeply implicated in the paradox of environmentalism. Unlike Bramwell, he believes that a democratic people are something more than greedy little pigs and that America's well-being does not rest solely on economics. Like Paehlke, Sagoff sees economic prosperity as only one of several, sometimes competing, first-order values that engender processes of discussion, negotiation, and compromise. Social regulation, Sagoff argues, "expresses what we believe, what we are, what we stand for as a nation, not simply what we wish to buy as individuals. Social regulation reflects public values we choose collectively, and these may conflict with wants and interests we pursue individually" (1988, 16–17). This argument underscores the importance of religion to environmentalism, if for no other reason than the possibility that religious discourse might expand our political conversation beyond economics. Discourse is the saving grace of democracy, and green politics—whatever else it might represent—entails a public discourse expanded beyond the language of utilitarian individualism.

Green Politics and the Debate over Ecology

Some commentators believe that the term *environmentalism* is itself a carryover of the DSM. Charlene Spretnak (1986), for example, argues that environmentalism per se is not what green politics is about. She claims that green politics involves a lot more than, for example, cleaning up the air and developing a national energy policy, in that it also includes a rejection of anthropocentrism (27), a denial of the mechanistic underpinnings of the industrial growth paradigm (29), and a renunciation of patriarchy, including the domination of nature and woman by vertical-hierarchal modes of institutional organization (30). "The spiritual dimension of Green politics, then, will have to be compatible with the cultural direction of Green thought: posthumanist, postmodern, and post-patriarchal. That direction will probably come to bear the inclusive label 'postmodern'—*unless* that tag has already been ruined almost before we have begun" (34). Spretnak's argument implies that green politics and environmentalism are synonymous only if conceived outside the frame of the DSM—a distinction sometimes made by contrasting *reform environmentalism* (roughly equivalent to progressive conservation) with *radical environmentalism*. Green politics for Spretnak heralds the possibility of a cultural paradigm shift, something more than normal conservationist and environmentalist politics.

Paehlke, in distinction from Spretnak, believes that the term *environmentalism* is today, whatever it meant in the past, more in line with the NSM than the DSM. He believes that the new or "second wave environmentalism," beginning during the 1980s, has expanded beyond traditional conservationist issues to include "urban issues such as pollution and energy sustainability" and has begun to question our blind faith in "economic growth and technological 'progress' "(1992, 3).[13] Environmentalism and green politics are virtually synonymous for Paehlke, since both involve "fundamental ideological, political, and institutional questions" (3). Since 1986 public attention has been captivated, Paehlke believes, by dramatic issues of global ecology, such as the destruction of tropical rainforests and holes in the stratospheric ozone layer. But more mundane problems, such as the cost of energy and the management of solid waste, have not escaped the public's attention either. Still, despite the visibility of environmental crisis and the public clamor for action, Paehlke continues, "only a limited number of people appreciate that environmental concerns provide a fundamental challenge to the structure (though not necessarily the market-based organization) of our economy, our political institutions, our policy-making habits, our cultural values, and, in a phrase, the North American way of life" (5). Like Spretnak, then, Paehlke does not argue that green politics has arrived, in the sense of achieving a social consensus, but he believes that environmentalism is becoming more influential in setting the social agenda.

More than anything else, Paehlke thinks of environmentalism as an emerging cultural paradigm. "Historians [such as Hays] as well as philosophers [such as George Sessions] have observed that the contemporary environmental movement is based on a transformation of human social values. . . . Whatever name one attaches to the change, the environmental movement is the political manifestation of a significant shift in societal values" (1990, 350–51). Some of the values Paehlke thinks are characteristic of North American environmentalists are listed in table 3. Obviously, not every environmentalist or environmental group holds each of these preferences.

Just as obviously, the existence of these preferences does nothing to ensure the election of environmentally concerned politicians, the passage of ecologically informed public policies that lead toward sustainability, and the institutionalization of these policies. The paradox of environmentalism implies that the environmental movement is not yet an effective political force, if for no other reason than neither political party is an ecological standard bearer. Politicians who advanced ecologically informed platforms, such as former Senator and now Vice President Al Gore, generally do so not because of but in spite of their political affiliations. Further, each of the first-order political values identified by Paehlke (national security, economic prosperity, social justice, and democracy) has a variety of constituencies that advance and defend it. Economic prosperity is defended by investors, trade unions, civic organizations, corporations, and consumers. Indeed, the political institutions of American society are dedicated to the first-order value of economic growth, embedded in the language of utilitarian individualism. Other primary values like

Table 3 Defining Values of Environmentalism

- An appreciation of all life-forms and the complexities of the web of life.
- A sense of humility regarding the human species in relation to other species and to the global ecosystem.
- A concern with the quality of human life and health.
- A global rather than nationalist viewpoint.
- A preference for political decentralization.
- A concern about the long-term future of planet Earth.
- A sense of urgency regarding the survival of life, especially in protecting biodiversity.
- A belief that human societies ought to be reestablished on a sustainable basis.
- A revulsion toward waste (shading into aestheticism in its extreme forms), especially with an emphasis on recycling.
- A love of simplicity (though not antitechnology in any generic sense).
- An aesthetic appreciation for nature and its processes.
- A belief that qualitative values contribute to self-esteem and social merit.
- An attraction to autonomy and self-management; an inclination to participatory political processes and administrative structures (citizen democracy).

Source: Adapted from Paehlke 1988

national defense are linked to economic prosperity. It is no accident that politically influential states like Texas and California are homes to the defense industry. And even issues of social justice, such as housing and health, have been viewed through the lens of utilitarian individualism. In distinction from all the other first-order values of our society, environmental politics is less dominated by material self-interest (although environmental contractors have reaped rewards).

Daniel Botkin offers yet a third definition of *environmentalism* as "the social, political, and ethical movements that concern the use of the environment" (1990, 230). This definition, however, includes virtually every "environmental" group, including groups with opposed goals. For example, on Botkin's definition the progressive conservation movement is part of environmentalism, since it (the yet dominant ideology in America concerning the use of the environment) involves social, political, and ethical goals. Even more problematically, so-called wise-use organizations, which propose to mine and log every available acre of America's public lands and forests, are included. And so is the Earth First! organization, though it presents an ideology antithetical to progressive conservation. Thus, Botkin is neither as sanguine as Spretnak that a "posthumanist, postmodern, and postpatriarchal" paradigm will soon emerge nor as optimistic as Paehlke that established environmental values are beginning to push America in a green direction.

Spretnak's, Paehlke's, and Botkin's definitions of environmentalism share one thing: all point toward the importance of a normative component in environmentalism. Yet the steady deterioration of the health of our ecosystem over the past two decades implies that environmentalism is more a mélange of competing interests than a well-defined social movement—that it lacks, as McCann argues, moral authority. Which is to say that no effective political consensus on an environmental agenda yet exists. Although Spretnak, Paehlke, and Botkin associate questions of ethics and values with environmentalism, most Americans have not yet considered the ethical issues that ecocrisis raises. Further, none of these definitions help define the role of science (biology, ecology, geography, and so on) in environmentalism. This question, as argued in chapter 1, cannot be answered apart from setting ecology in a cultural context that includes religion. Botkin and Paehlke envision an expanded role for ecology in environmentalism, that is, in determining public policy and thus cultural outcomes. Paehlke argues that environmentalism should be viewed as a social movement that is (at least in part) inspired by an ecological paradigm. If environmentalism is to be successful, he continues, "it must include a set of values and claims regarding the role of science in society" (1989, 143). Yet Bramwell, who believes that ecology is in fact influencing the public agenda, perceives ecologism negatively—a threat to Western civilization.

The emphasis on a science component in environmentalism is nothing new. Worster ([1977] 1985) points out that Americans have expected ecology to contribute to solving environmental problems at least since the Dust Bowl of the 1930s. Ecology generally has been either incapable or unwilling to assume this function. Economics, as Robert P. McIntosh (1985) points out, still dominates public policy-making. One reason is that economic growth is socially legitimated and institutionalized in the American system of governance. Environmental values, in contrast to economic ones, are not legitimated by consensus. And ecology has not articulated a paradigm that integrates scientific judgment with public policy. Ecology also lacks influence, according to McIntosh, because it "does not fit readily into the familiar mode of science erected on the model of classical physics, and it deals with phenomena which frequently touch very close to the quick of human sensibilities, including aesthetics, morality, ethics, and, even worse in some minds, economics" (1). And, quoting L. K. Caldwell, McIntosh notes that ecologists have ignored the potential of their discipline to mold public policy since they have "been focused largely upon manageable microproblems from which the human animal was usually excluded" (313). Environmental science has, in effect, been caught between two competing images—objective and normative.

Barry Commoner's book *The Closing Circle* perhaps exemplifies a *normative ecology*. He argues that ecology establishes four laws that humankind is ethically obligated to follow: everything is connected to everything else; everything goes somewhere; there is no free lunch; and nature knows best. According to Commoner, a green society can be created simply by following "nature's ways." Survival as a culture, he concludes, requires that "ecological considerations must guide economic and political ones" (1971, 291). In chapter 1 I noted the naïveté of such a view, since it ignores the bonds between science and culture. Science is embedded in an environment that gives it direction, purpose, and meaning. It would seem that if there were any quick passage from ecological science to sustainability, to a better way of life not only for this generation but for all those to come, society would have taken that path. The paradox of environmentalism confirms that no rapid reorientation of the industrial growth society is occurring. As McCann notes, the problem is that although ecology "can help demarcate the circle of contingencies and risk probability that constrains our options, it provides little instruction or justification as to the human purposes for which we should act within those constraints" (1986, 248).

Botkin is far less sanguine than Commoner about the prospects of ecology to guide society toward sustainability. Though he does not take Commoner specifically to task, Botkin challenges the idea of a normative ecology and argues against the idea that natural ecosystems are more resistant to collapse than ones simplified by human action. Unhumanized ecosystems, he maintains, simply do not have the stability and order that environmentalists characteristically assume. From his perspective the ecological facts are less straightforward than some scientists imply. How can we know which face of nature to follow since it sends so many divergent signals? *Discordant Harmonies* seems to reach conclusions not altogether incompatible with those of Spretnak and

Milbrath—namely, that achieving sustainability requires a "change in metaphor, myth, and assumption" as much as or perhaps even more than the facts themselves (vii).

Even Aldo Leopold's land ethic ([1949] 1970) is subject to the kind of criticism that Botkin makes. Leopold argues that economics has assumed the upper hand in American life, and as a result we are undercutting the fundamental biological processes that sustain culture. Leopold's land ethic, however, offers a new point of departure quite distinct from the governing ethic of resource conservation. That tradition, established by President Theodore Roosevelt and Gifford Pinchot at the turn of the century, sees nature through utilitarian spectacles, as little more than a stockpile of natural resources for human exploitation. Leopold, in contrast, advocates that we stop thinking about "land-use as solely an economic problem. Examine each question in terms of what is ethically and esthetically right, as well as what is economically expedient. A thing is right when it tends to preserve the integrity, stability, and beauty of the biotic community. It is wrong when it tends otherwise" (262). He believes that science can help move society toward sustainability because individual citizens, armed with ecological knowledge, would make the right choices. Botkin's arguments imply that ecology cannot provide such a guide, if for no other reason than the lack of any knowledge of general ecological laws.

McIntosh is skeptical that ecology can contribute much to public policymaking for another reason—that it cannot be easily integrated with social issues. He points out that, almost coincidentally with Earth Day One (April 1970), ecologists called for the creation of a National Institute of Ecology to integrate ecology with public policy. "This effort to integrate ecology with general public policy was, if not premature, apparently beyond the scope of the capabilities of ecology as an 'integrative science'" (1985, 312). Part of the problem is that ecology remains in some ways in a preparadigmatic state; no one set of basic assumptions governs ecological science. McIntosh claims that ecologists are divided over the relevance of temporal variables to ecological inquiry (6). Does ecology seek knowledge of time-independent (ergodic) phenomena? Or does it seek knowledge of time-dependent (historical) phenomena? Because the community of ecological science remains diverse, what sociologists of science call "normal science"—where the vast majority of practitioners function within a widely accepted set of conventions—is not yet possible (318). The preparadigmatic nature of ecology is perhaps exemplified by the split between so-called shallow and deep ecology, a split that virtually mirrors the gap between progressive conservationists and radical environmentalists. An even more fundamental difficulty, however, is the lack of a social consensus to enable environmental science to assume a greater role in shaping public policy. As McIntosh suggests, ecology is a reluctant science vis-à-vis public policy, in part because the strength of ecology is its science component, that is, its quantitative and not its qualitative or ethical side (319).

However, as the Ecological Society of America acknowledges in its Sustainable Biosphere Initiative (SBI), part of the problem is that there is no public mandate to legitimate the use of ecology in setting public policy. Issues like integrity and ecosystem health involve normative considerations. As Levin argues, not only scientific but sociological (that is, political, ethical) considerations are involved even in defining ecological indicators. "Concepts of sustainability inevitably reflect societal choices at the local, regional, and global levels. Balancing ecological with economic and other considerations is the fundamental problem in defining sustainability" (1992, 213). Although science can inform the policy-making process, Levin argues that scientists "cannot usurp the role of humans in . . . decision-making involving multiple-use systems. Partnerships must be found between the scientific process and the policy process, or else decisions will continue to be made without adequate scientific input" (213). He also notes that the SBI attempts to counter the tradition among scientists that relies on individual investigators and their agendas. Ecology as a profession, he argues, "cannot carry out business as usual: the urgency of global environmental problems demands that research be performed to address them" (213). Which completes the circle of my argument: religion is the most plausible means of legitimating ecology's role in setting public policy and of democratically empowering a movement toward sustainability.

Religion and Environmentalism

A recapitulation may be useful. I have argued that the environmental crisis threatens to irreparably harm the biophysical processes that sustain life on earth. I have also claimed that religion is involved in devising solutions for ecocrisis because, in part, the scientific information necessary to inform the policy-making process can be incorporated only via an empowering political matrix that presupposes solidarity on the issue of caring for creation. The paradox of environmentalism confirms that the philosophy of managing planet Earth dominates our culture. *But great cultural crises are always moral crises.* Religion is the most likely way that Americans can move themselves to care for the Creation. If an effective social consensus can be created, the problem of a "reluctant ecology" would be resolved. Further, Botkin's arguments against normative ecology can be reinterpreted as shifting the role of ecology in guiding societal decision-making to its proper domain: the cultural matrix. Botkin's argument is thus consistent with *The Economy of the Earth*, in which Sagoff argues that environmental policy is properly understood as a political and ultimately ethical question having to do with America's identity as a nation rather than with Americans' individual choices as consumers. According to Sagoff, no economic method can either substitute for or replace the role of citizens—even if that role is indirect, that is, exercised through the election of environmentally conscious politicians—in determining environmental policy choice. If citizens default to economic experts, policy choice is restricted to a narrow monetary calculus of value.

An almost exclusively and narrowly defined economic decision-making matrix has led America into ecocrisis: nature is not simply, though economists like to think so, a "cash cow" (see Georgescu-Roegen 1971 and Daly 1991). The alternative is a citizen democracy, which involves open-ended conversation and deliberation about social preferences that lie outside the market. To date, environmental issues have been viewed within the dominant framework of cost-benefit analysis, largely because (as Dahl implies) associational pluralism restricts environmental discourse to the politics of interest. The politics of interest is aided by the suspicion with which the American public regards government and by the notion that economic growth and progress are synonymous. Notwithstanding the perhaps 80 percent of American voters who think that "environmental standards cannot be too high" and that improvements must be made "regardless of cost," *there is no consensus of values that can undergird political action and institutional change.*

Several arguments support the thesis that religious discourse is essential to shaping and refining environmentalism as a social, moral, and political force that, at a minimum, might allow the United States to better the more onerous and threatening aspects of ecocrisis and at best might move society toward sustainability.

The church presents the readiest opportunity for most Americans to engage in a discourse concerning the public good. In this regard, Canada and especially the United States are exceptions among the West's democratic societies because organized religion continues to flourish in these countries. More than 90 percent of Americans believe in God, divided among 1,200 denominations with nearly 400,000 congregations. Of the faithful, nearly 25 percent are Roman-Catholics; 24 percent are moderate Protestants; 16 percent are conservative Protestants; 9 percent are liberal Protestants; 9 percent are black Protestants; 8 percent belong to other faiths; 7 percent have no religious preference; and 2 percent are Jewish. Almost two-thirds of these religious Americans characterize themselves as actively involved in their congregation.

In chapter 5, I develop a detailed description of how the church provides opportunities for ordinary people to inform themselves about environmental issues in a religious context, begin to converse about the good society, and ultimately take action with local, national, and even global consequence. The essential point here is that *the church is a mediating institution*, closer to us as individual human beings and as moral agents than the more powerful institutions of business and politics. As a mediating institution, the church can help close the gap between private values, the utilitarian individualism that poses as a public philosophy (what we desire as consumers), and public values, which signify what we are and hope to be as a people, as a culture (what we desire as citizens). I agree with Doug Bowman that the greatest service the church can render at this time is to provide "education concerning the facts of the environmental crisis, the ingredients in our current thinking and action that contribute to the crisis, and the resources that we already have available to us for altering our way of thinking and action in constructive ways" (1990, 86–87).

Public values have become an endangered species in the twentieth century, whipsawed between the interests of giant, often multinational corporations and big government, and the complexities of the issues, whether these be defense, welfare, or ecology. What environmental legislation has been enacted has forced a continuous struggle—one that has met with increasing opposition, as Samuel P. Hays argues, during the late 1970s and throughout the 1980s. The opposition, in fact, has become "a coherent antienvironmental movement" whose objective is that of "restraining environmental political influence" (Hays 1987, 287). During the 1990s, big government and corporate interests have appeared to environmentalists to act in concert against the green agenda. Former Vice President Dan Quayle's Council on Competitiveness, which had the power to override environmental legislation in order to foster economic growth and production, was perhaps the most notorious example. In early 1992 the council issued an edict permitting violations of the clean air act *without public notice*. Profit at the expense of the commons and public health, in other words, took precedence over the rights of government by, for, and of the people.

The need for the church to serve as a mediating institution in relation to the corporation is readily apparent: corporate decision-making is insulated from democratic process. Corporations are publicly chartered but privately owned and managed entities that are oriented primarily toward the bottom line.[14] Of course, some corporations go beyond the bottom line, but they are the exception. Private interest, not the public good, is necessarily the limit of corporate decision-making. Corporate management is usually a self-perpetuating oligarchy in which success, by definition, promotes private interest. The corporate culture essentially rewards those who contribute favorably to the bottom line, which entails a perspective limited to a short-term economic horizon. Of course, corporations are concerned with so-called public relations. But "public relations" does not involve democratic process, the inclusion of the public in discretionary decision-making, but it is rather the endeavor to develop a favorable image, partly to head off legislation that would restrict corporate activity, and partly to avoid having to change the corporate status quo.

Some political scientists, such as Dahl, raise questions about the legitimacy of corporate oligarchy. Abandoning the standard economic interpretation of corporations (the argument that no reason exists to extend democratic process to a corporation's internal politics since no government exists therein), he argues that since the existence and operation of corporations necessarily require "relations of power and authority," then as members of a democratic society "we are entitled to ask—indeed we are obligated to ask—how that government should be constituted" (1989, 329). A variety of proposals to make corporations more responsive to the public interest, such as C. D. Stone's (1975) idea that public representatives be added to corporate boards of directors, have been made. Stone also argues that laws like the National Environmental Policy Act must be either amended or supplemented to require that "companies of a certain class *must* establish vice-presidents for environmental affairs" (124). Other commentators, among them E. E. Spitler (1992), argue that criminal penalties for corporate officials who fail to ensure compliance with environmental law are required. But no such reforms will occur until the majority mandates change by electing politicians who support enabling legislation.

The church can also serve as a mediating institution in relation to big government. Politicians and government bureaucrats march to the first law of politics: survival above all else. Because political survival depends in America on being able to attract donations to finance campaigns, special interest groups have enormous influence on the political process.[15] Daniel Koshland (editor of *Science*) observes that the half-life of a politician is four years, plus or minus two. Issues such as the survival of species or ecosystems require a long-term perspective that politicians simply ignore. Politicians thus respond to the interests of present voters, and economic issues take precedence over all others (Koshland 1991). As Johnston points out, "Governments are elected for limited lives. . . . Thus anything but a relatively short-term future is of little relevance" (1989, 170). Politicians want, above all else, to be re-elected.

The local church, though it provides no political panacea, offers a chance for citizens to participate in democratic debate over environmental issues in substantive ways outside the campaign process and in ways that expand the deliberative agenda beyond issues of the moment. As a mediating institution, apart from government and politics as usual, the church provides a forum for discussions that involve more than empty rhetoric, glittering generalizations, and slogans that masquerade as public policy. Electoral politics has become consumed with form rather than substance, with personality rather than issues. Presidential candidates position themselves carefully on the issues, letting slogans ("I am the environmental president") and empty generalities ("We've taken decisive steps to eliminate air pollution") replace systematic discussion of the issues and policies that address them. The "sound bite" has replaced genuine debate in campaigns; information is unscrupulously manipulated by candidates; promises are made but seldom kept. On the administrative side of government, "policy analysis" supplants democratic discussion, which implies, as Bellah puts it, that we are coming perilously close to abandoning "the democratic undertaking altogether, and to . . . [admitting] that we have become the administered society our prophets have long feared we might become" (1991, 306).

In short, big government today excludes ordinary citizens from influencing decisions.[16] Al Gore argues in *Earth in the Balance* that "political motives and government policies have helped to create the [environmental] crisis and now frustrate the solutions we need" (1992, 11). From his perspective, the paradox of environmentalism reflects the reality that "current public discourse is focused on the shortest of short-term values and encourages the American people to join us politicians in avoiding the most important issues and postponing the really difficult choices" (11). While environmentalists have welcomed the arrival of the Clinton-Gore administration and believe that, for example, the EPA will actually begin to enforce the law of the land, none have great hope that the United States is moving toward fundamental environmental reform, that is, toward sustainability. Politicians, we must remember, are followers, not leaders: the greening of Clinton-Gore depends upon the greening of society. The philosophy that we can manage planet Earth will probably continue to govern our society. The course of events in the 1992 campaign bear this out: as the Clinton-Gore ticket was repeatedly attacked by the Bush-Quayle campaign for environmental extremism, Gore's environmentalism became increasingly muted and almost disappeared. Clinton-Gore strategists note that three issues—economy, economy, economy—provided the focal point of the Clinton-Gore campaign. Environmentalists believe that many of the innovative plans outlined in Gore's *Earth in the Balance*, such as a carbon tax designed to reduce greenhouse gases, have virtually no chance of being implemented.

As the Clinton presidency illustrates, the question that confronts environmentalism is how to promote discourse that encourages widespread citizen involvement, deals with the ethical issues posed by ecocrisis, and politically empowers a social movement toward sustainability. The church, I am arguing, is the most likely place for this to happen, since every tradition of faith has resources to support an environmental ethic that cares for the Creation. Which leads, then, to the second argument for the relevance of religious discourse to environmentalism:

Religion has been throughout history and remains today the central source of criticism and resistance to the state. Environmentalism involves more citizen democracy, that is, participatory political processes and administrative structures, and less authoritarian control, expert planning, and top-down management (see tables 2 and 3, above). Compared with most of the world's political systems, American democracy is ideal. But we must not let our commitment to the ideals of democratic life (polyarchy, as Dahl terms it) blind us to the reality of the state: its overriding purpose is to serve industrial growth, the dominant social matrix that is driving the earth toward ecocatastrophe. In his critique of the modern age, Bowman chides Americans for their idealistic "assumption that somehow the continuation of current policies will bring us through the period of the inexorable destruction of the biosphere. But in fact it must be recognized that these policies and the way of thinking they are founded upon are bankrupt" (1990, 78). At least we should rethink the relation between economics and ecology.

Environmentalism, conceptualized as a sociopolitical movement that aims to ameliorate ecocrisis by restoring citizen democracy, coincides with the historical stream of Judeo-Christianity. Thousands of years ago, the Yahwist religion rose up as a revolutionary force to challenge the political and military hegemony of the hieratic civilizations of the Near East. As Norman Gottwald argues in the *Tribes of Yahweh*, Yahwism opposed the oppressive, socially stratified states that surrounded Israel. Herbert Schneidau (1976) contends that the

Yahwists defined themselves through a critical opposition to the Sumerian and Egyptian systems of belief. Similarly, the early Christians were at odds with the political power of the world—Rome itself. The persecution of Christians under the Roman Empire is legendary.

Examples of the potency of Judeo-Christianity as a political force in the United States can also be cited. Garry Wills (1990) contributes case studies ranging from the Civil War through civil rights to abortion. He argues that Lincoln's call for abolition is incomprehensible apart from the biblical tradition in which he was steeped. The great war, ultimately, was a test of faith—of the people's ability to work out God's will, the "very ordeal . . . a sign of God's superintendence, bringing good out of evil" (218). Similarly, the civil rights movement has firm roots in biblical tradition, vividly represented by Martin Luther King's appeal to natural law in his classic "Letter from a Birmingham Jail." Another example is the Vietnam War protest movement, perhaps the most massive episode of religiously inspired civil disobedience in American history. Clearly, the church has risen up time and again in opposition to the state and its policies: political crises, as Wills observes, are often moral crises. And insofar as any political crisis involves a moral dimension there is no solution independent of considerations of the meaning of life and questions of value.

Religious discourse is essential to shaping and refining environmentalism for a third reason: the notion of justice. No form of government is intrinsically just; justice is an achievement enabled through political practice. Of course, historical and cross-cultural comparison give evidence that some forms of governance, for example, democratic forms, significantly increase the possibility for justice. But even a democratic society is not intrinsically just. As Abraham Kaplan puts it, "In America the force of the state does not define morality but is itself continually subjected to the moral judgment of the citizen" (1963, 70). This relation is in fact the essential underpinning of a democratic state, since the citizenry supports the state not at the point of a gun but because of an understanding that the state is moral. The idea of moral judgment leads to important distinctions between procedural justice, distributive justice, and substantive justice.

America receives high marks for its system of due process, that is, *procedural justice*, which ensures that its laws are enforced and applied fairly and that legal remedies are (almost) equally available to all. Of course, powerful corporations and wealthy individuals have advantages over relatively less powerful public interest groups and impecunious individuals in seeking legal remedies.

Although environmental advocates have, for example, brought class-action suits against corporate polluters, corporations have counterattacked such individuals, involving them in often ruinous counter-suits for libel and damage. On balance, much of the progress that has been made in resolving ecocrisis has been through the court system. Still, the paradox of environmentalism immediately reminds us that the presently existing system of law and precedent is not up to the challenge of preventing the relentless degradation of the earth.

The United States has also achieved a relatively good record in *distributive justice*. No one can claim that this nation is a socioeconomic utopia, but in the main only individuals who are educationally disadvantaged (itself a result of a lack of substantive justice) are precluded from enjoying economically adequate ways of life (and associated benefits, such as health care).[17] Peter Wenz's "concentric circles" theory of distributive justice (1988) offers an interesting approach to environmental issues. His examples, ranging from agriculture to energy policy, illustrate how society might begin to address ecocrisis through deliberations over distributive justice. Wenz argues that the present system of agricultural production, for example, treats future generations unjustly, because they "are going to have to pay part of our food bill" due to the loss of prime agricultural land to population growth and declining soil fertility (333). What is the remedy? Governmental intervention, he suggests, is needed, most likely in the form of "subsidies for traditional farming" that "increase the current generation's total expenditure on food production so that future generations will not have to bear the cost of feeding us" (333).

Substantive justice concerns what might be called the final ends of a people: its conception of a just, fair, and good society. Considerations of substantive justice are crucial to the success of environmentalism. Wenz argues that any perception by the public that policies "are consistently biased in favor of some groups and against others could undermine the voluntary cooperation that is necessary for the maintenance of social order" (1988, 21). For a democratic people to act, as in the case of redirecting society toward sustainability, people need to believe that the proposed actions are just. Thus, Wenz concludes, "environmental public policies will have to embody principles of environmental justice that the vast majority of people consider reasonable" (21). However, he places faith, perhaps too much, in principles and theories of justice to motivate people. Religious discourse might be more effective than Wenz's philosophical discourse in moving society toward sustainability, and in any case, religious narrative often converges on questions of substantive justice.

Which is to say that religious discourse is a second language—the Great Code itself, the legitimating narrative that underpins the West—that might enable a democratic citizenry to overcome its tendency to think of the common good only in terms of the first language of utilitarian individualism. No cogent argument can be based on the notion that we are *Homo sapiens*, rational animals, who will use "reason," as in the appeal to Wenz's concentric circles theory of justice, to solve ecocrisis. Consider the following argument:

> If we were rational animals, then we would not be systematically destroying our environment. We are destroying the environment. Therefore, we are not (at least not totally) rational animals.

At this juncture, it should be apparent that the modern paradigm (or DSM) conditions a kind of rationality (profit maximizing, cost minimizing) that destroys the environment. The system of categories and theories in terms of which we discriminate the things in the world and the relations among these things are "reason at work," an ideological system that rationally degrades the biophysical processes of life on earth. But there are alternative story traditions, outside the DSM, which might lead a democratic society toward sustainability—in ways beyond philosophical arguments for environmental justice. The biblical tradition undergirds the largest (most extensive) and most effective community of memory operating in American society, a legitimating narrative that remains unencumbered by the modern story, that is, the narrative of utopian capitalism. Chapter 3 discusses how the biblical tradition provides a normative backdrop against which questions of environmentally oriented institutional reform and policy choice might occur.

3

The Sacred Canopy: Religion as Legitimating Narrative

Every human society, however legitimated, must maintain its solidarity in the face of chaos. Religiously legitimated solidarity brings this fundamental sociological fact into sharper focus. The world of sacred order, by virtue of being an ongoing human production, is ongoingly confronted with the disordering forces of human existence in time.—Peter Berger, *The Sacred Canopy*

A startling reality is revealed when religion is viewed from a sociolinguistic perspective. Religion, or more accurately the rituals, symbols, institutions, and texts through which religion is objectified, has functioned throughout history to legitimate the social order—that is, to provide a context of ultimate meaning. Peter Berger argues that religion makes society possible by providing "a sacred cosmos that will be capable of maintaining itself in the ever-present face of chaos" (1967, 51). There is good reason to think, even amid the secularism of the late twentieth century, that religious discourse provides a sacred canopy for the vast majority of Americans. More specifically, the overarching context of our fundamental beliefs—that time is meaningful, that human life is morally significant, and that humans can live together in a good society—have been and are being shaped by the Judeo-Christian narrative tradition.

Secularists who believe that religion is a dead letter will be dubious about such an assertion. This skepticism is ironic, since a wide variety of secular scholars, including both natural and social scientists, confirm that religion overdetermines us. My sociolinguistic perspective on religion as providing a sacred canopy, a telic structure that subtly (even unconsciously) establishes the basic possibilities of human beingness, is reinforced by a number of theoretical perspectives, including cultural anthropology and sociobiology. These disciplines advance the premise that human behavior is biologically underdetermined. Which is to say that the human gene pool alone is incapable of either providing for or explaining the reality of specifically *human* life. As the cultural anthropologist Clifford Geertz puts the point, it is "because human behavior is so loosely determined by intrinsic sources of information," that is, the genes themselves, "that extrinsic sources are so vital" (1973, 93). By extrinsic sources Geertz means symbolic patterns or systems of symbols, like religious discourse, that "lie outside the boundaries of the individual organism" (92).

From this perspective, the most interesting (unusual) things about human behavior are not its biological but its cultural determinants. Analogically, Geertz continues, just "as the order of bases in a strand of DNA forms a coded program, a set of instructions . . . for the synthesis of the structurally complex proteins which shape organic functioning, so culture patterns provide such programs for the institution of the social and psychological processes which shape public behavior" (1973, 92). Though genes and symbols are analogous, however, there are important differences. Geertz suggests that genes are *models of* an observable pattern or order but that symbol systems (such as a theology) are not only *models of* but also *models for* observable patterns or orders: "they give meaning, that is, objective conceptual form, to social and psychological reality both by shaping themselves to it and by shaping it to themselves" (93).

Grasping the notion that the human species is biologically underdetermined and culturally overdetermined places religion in an appropriate context.[1] Religion, comprehended as symbolic discourse, has been and remains a nearly universal phenomenon that helps to direct (organize, guide) human behavior—that is, it provides instructions for the expression of genetic potentials. Geertz offers a useful definition of religion consistent with a sociolinguistic orientation. Religion is "(1) a system of symbols which acts to (2) establish powerful, pervasive, and long-lasting moods and motivations in men [and women] by (3) formulating conceptions of a general order of existence and (4) clothing these conceptions with such an aura of factuality that (5) the moods and motivations seem uniquely realistic" (1973, 90).

The sociobiologist Richard Dawkins helps us grasp Geertz's definition by describing the vehicle of cultural transmission as the *meme*. Memes depend on language and religious discourse is no exception. Cultural memes, religious or otherwise, perpetuate themselves through socialization, roughly analogous to the propagation of biological genes through reproduction.[2] But cultural transmission can also "give rise to a form of evolution. . . . Language seems to 'evolve' by non-genetic means, and at a rate which is orders of magnitude faster than genetic evolution" (1976, 203). Religion is a prime example of how the basic code for specifically human life is transmitted from generation to generation as well as how the code evolves in response to changing circumstances of existence. "Consider the idea of God," Dawkins writes. "How does it replicate itself? By the spoken and written word, aided by great music and great art" (207). From a sociobiological perspective God is a meme: both a living structure around which human life is organized and a legitimating narrative that directs life by giving answers to its most basic questions. "God exists," Dawkins continues, "if only in the form of a meme with high survival value, or infective power, in the environment provided by human culture" (207). Religion, then, can be seen as a co-adapted complex of religious memes organized around a God-meme (or, in Geertz's terms, religion is a cultural system).

Read in sociolinguistic terms, there is new meaning in biblical criticism, as in Harold Bloom's contention that "the primal author J [the so-called Yahwist, author of portions of the Old Testament] . . . constitutes a difference that has made an overwhelming difference, overdetermining all of us—Jew, Christian, Muslim, and secularist" (1989, 3). Similarly, Northrop Frye's (1981) argument that Western culture *remains* fundamentally a biblical culture assumes new meaning. The Bible is an exemplary text, a means by which Judeo-Christian culture maintains itself.

For theist and atheist alike many of the underlying principles and values that define life are rooted in the biblical tradition. Which is to say that the Great Code undergirds culture to the extent that our lives are incomprehensible without reference to the Bible. Without that linkage there can be no cogent explanation of how Western civilization has come to be. Nor, insofar as we are storytelling culture-dwellers, can we envision a future apart from temporal narrative. The notion of time as going somewhere is inexplicable outside the biblical tradition, for the idea that past-present-future signifies something rather than nothing is an invention of the Hebrew Bible. The creation stories and the covenant chronicles of the Old Testament create a temporal narrative, underscoring the notion that time is a meaningful linear and nonrepetitive passage of events. Consider the first verse of Genesis. "In the beginning God created the heavens and the earth." We moderns take "the beginning" for granted, as if such a sense was the human norm. It is not, even though temporal progression has a normative status for us. The covenant amplifies the sense of temporal order, as in Genesis 12. "Now the Lord said to Abram, 'Go from your country and your kindred and your father's house to the land that I will show you. And I will make of you a great nation.' " The covenant is part of God's plan, and the Hebrews are called to place their faith in it. This amplifies the Hebrews' sense of time, linking their lives with a sacred temporal order, which explains the intense consciousness of history, and therefore the significance of linear time, so characteristic of Judeo-Christianity.[3] Our myopia about the source of our sense of temporality, the taken-for-grantedness of the reality of time, testifies to the great and lasting legacy of the Hebrew Bible. We cannot conceive of our lives as being meaningful outside time's sacred canopy.

Once the vital role of culture in determining human behavior is grasped, then the diversity of the forms of human existence is brought to hand as well. Simply stated, the controlling memes of, for example, Swahili culture and that of twentieth-century North America are different. Religious discourse thus occupies, any scientific pretense to the contrary, a position of inordinate importance even within a culture like ours, where the legitimating narratives of modernity (economic, scientific, political) tend to obscure the sacred canopy. Such a perspective assumes relevance in the present context, for it begins to fill in the outlines of the thesis that environmentalism turns on religion. Religious belief—its symbols, rituals, and texts—offers possibilities to challenge the economic orthodoxy that nature is nothing more than resource as well as the political orthodoxy that the state is inherently moral.

Judeo-Christianity in Sociobiological Context

Cultural evolution is undoubtedly something different from biological evolution. Just as clearly, biological evolution underlies the possibility of cultural evolution. "The old gene-selected evolution," as Dawkins puts it, "by making brains, provided the 'soup' in which the first memes arose. Once self-copying memes had arisen, their own, much faster, kind of evolution took off" (1976, 208). My purpose is not served by attempting to unweave either the complexities of sociobiology or the debate over its merits.[4] The point is that sociobiology, as noted in chapter 1, presents an array of useful possibilities for explaining the relevance of religious discourse to ecocrisis. Loyal Rue presents one of these in his provocative book, *Amythia*. He traces the emergence of symbolic systems through history to the point at which the human project arrived at language, and thus metaphorical and abstract thought. Following Dawkins, Rue claims that "the integration of discrete neural systems in humans is orchestrated by the features of a verbal system" (1989, 39). This premise leads him back to the central concept of cultural memes, viewed as *metaphorical primes*—the fundamental units of meaning that determine most if not all information processing and reception, storage, and transmission.[5]

From Rue's sociobiological vantage point, Judeo-Christian mythology integrates the basic memes that define Western culture. Rue defines "myth as the *achievement* of an integration of cosmos and ethos, an achievement without which there would be no possibility of achieving personal integrity or social coherence" (1989, 46). Religious discourse enables the sacred canopy, the overarching structure that grounds human beingness in a meaningful cosmos. Central to the West's Judeo-Christian mythology has been the root metaphor of God-as-person.[6] Crucially, on the traditional Judeo-Christian account, the human being has been created *imago Dei*—in the image of God—thus integrating the cosmic, sacred order with the secular, worldly order of events. The Old Testament language, such as Genesis 1.26 ("Then God said, 'Let us make man in our image, after our likeness . . . .' So God created man in his own image, in the image of God he created him; male and female he created them") is intensified by the New Testament, for Jesus is both man and God. Consider, for example, Matthew 1.21. " 'Behold, a virgin shall conceive and bear a son, and his name shall be called Emmanuel' (which means, God is with us)." To put the point more directly, through the notion of God-as-person and its subsequent development in the Bible, human personality is set in and tied directly to a cosmic frame of reference.

In his book, Rue deals specifically with what he believes is Western culture's loss of faith in its metaphorical prime. There are Judeo-Christians who have not lost faith in God-as-person. Rue argues, however, that the advance of science and the increasingly secular nature of American society makes such a belief intellectually untenable for the majority of the faithful. As he sees it, the loss of faith explains why our culture is experiencing crisis. And why, also, we seem collectively incapable of acting even when confronted by the possibility of ecocatastrophe. A *viable myth*, he argues, grounds ethos in cosmos, some larger frame of reference than human purpose alone. Above all else, Rue continues, the governing *mythos* must be a shared vision. Otherwise culture runs an increasing risk of dissolution and collapse.

According to Rue, contemporary science gives us an evolutionary picture of the cosmos that contradicts the metaphorical prime of Judeo-Christianity, that is, a transcendent, creator God-as-person. As a result, most people are incapable of believing in the basic mythos of the West. And therefore we are psychologically incapable of acting collectively. This argument undergirds Rue's contention that any mythology adequate to escaping from amythia must possess both scientific plausibility (especially that it be true, that it have adaptive significance) and religious distinctiveness (since religion lies at the heart of the selection process of culturally controlling memes). Or, to rephrase the matter in Geertz's terms, when we lose faith in the *model for* reality, culture drifts. People lack motivation to deal with the circumstances of existence because they lack a governing mythos. Or, to consider yet a third account, that of Stephen Toulmin, the modern world has lost coherence between fact and value or between the actual order of things and the ideal order of things. The modern "world view of Descartes and Newton no longer represents a genuine *cosmos*. Instead, it is split down the middle" (1982, 224). Restated in sociolinguistic terms, the sacred canopy is in tatters.

But there is a *crucially important distinction* between Rue's argument and my own. Rue believes that the metaphorical prime for a postmodern society will be drawn from evolutionary science and melded with Judeo-Christianity. As pointed out in chapter 2, accounting for the role of religion in a time of ecocrisis does not necessitate such an expansive argument. My argument depends on denying neither any of the divergent claims to ultimate knowledge nor the metaphysically divergent accounts of the Creation. Indeed, divergent claims to ultimate knowledge of the Creation enable a variety of stories about caring for creation that can collectively create solidarity and move us democratically toward sustainability. *Caring for Creation* does not argue for the truth of evolution as a metaphorical prime in preference to the Judeo-Christian orthodoxy of God-as-person, at least in part because such a truth claim brings scientific and religious narrative into opposition. As I have already argued, in relation to ecocrisis it is more constructive to view religious and scientific narrative as collaborative rather than antagonistic. Judeo-Christians metaphysically committed to the God-who-is-there can, I claim, care for the Creation just as much, and perhaps more so, than people whose creation story is an evolutionary one.

Judeo-Christianity and the Language of Survival

Many scholars argue that Judeo-Christianity (as a covenant tradition) has always been bound up with the language of survival. The question is whether Judeo-Christian faith in its many varieties—or any alternative—can meet the contemporaneous challenge to survival. So placed in context, the issue posed by ecocrisis is more a pragmatic decision than one of determining the one right metaphysical interpretation of Judeo-Christianity, though I acknowledge that such theological arguments are important to believers. Every tradition of faith offers toeholds, consistent with and growing out of whatever claims to ultimate knowledge that faith makes, for determining useful responses to novel circumstances. As Norman Gottwald argues, "efforts to draw 'religious inspiration' or 'biblical values' from the early Israelite [and Christian] heritage will be romantic and utopian unless resolutely correlated to both the ancient and the contemporary cultural-material and social-organizational foundations. . . . The religious symbolism for such a project will have to grow out of an accurate scientific understanding of the actual material conditions we face" (1979, 706). Systems of religious practice and belief, he continues, "claiming to be based on 'biblical faith' will be judged by whether they actually clarify the range and contours of exercisable freedom within the context of the unfolding social process" (708).

In chapter 4, I shall show how religion offers a powerful language for dealing with ecological crisis, especially through ethical narrative that turns on the metaphor of caring for creation. Here I provide a brief example to confirm the possibility of caring for creation. My point is to illustrate, through a vivid contrast of two highly respected scientists with long experience in rainforest research, that an ethical commitment to care for creation can grow out of divergent, even contradictory, creation stories. One is E. O. Wilson, a noted American scientist with whom we are already familiar; the reader will recall how much weight my argument places on his research. The other is Ghillean T. Prance, a scientist who is well known in professional circles though less familiar to the North American lay-public than Wilson.[7] Prance, whose specialty is ethnobotany, is director of the Royal Botanic Gardens at Kew.

Metaphysically construed, Prance's and Wilson's cosmological models are irreconcilable. Prance is a conservative Christian, Wilson a scientific materialist. For Prance fundamental truths about the Creation are laid down in Genesis; for Wilson the Genesis creation story is mythic, not scientific. Yet construed as legitimating narratives, both positions enable a strong conservation ethic. Prance (1992) argues for a stewardship of Creation that is derived from biblical teachings (for example, Job 12.7–9) and is ecologically informed. Wilson (1992) argues for an environmental ethic that, like Prance's notion of stewardship, is ecologically informed. Wilson's environmental ethic (1984, 1992), however, derives not from biblical teachings but from the reality of co-evolution, prudence, the love of life (biophilia), and the idea of wilderness.

The contrast between Prance and Wilson is instructive, for it confirms (at least in a preliminary way) the possibility that science and religion are not necessarily antagonistic and that a practical environmental ethic is not only consistent with, but can also be derived from, religious sources. The implication in terms of democratic theory is promising. For most Americans, I have already claimed, the route to an environmental ethic is neither through any deep and abiding commitment to evolutionary theory nor to the philosophical systems created by environmental ethicists. It is through their religion. But faith does not require ecological illiteracy; indeed, no environmental ethic that is ecologically uninformed is practical in the present circumstances. Neither does an operative environmental ethic require Americans to abandon their present faith in favor of another "greener" one.

Change, especially religious innovation, is never easy. The power of the old and established to resist the new and tentative, even when change is related to survival, is well documented. Arnold Toynbee's *Study of History* reminds us that most civilizations collapse not because of alien invaders but rather through internal insufficiencies: they remain fixed in ways that are outmoded by the continuing stream of life. A Judeo-Christian faith that is oblivious to the circumstances of human existence within the web of life is a dead letter. Dawkins notes that co-adapted meme-complexes are selfish, exploiting "their cultural environment to their own advantage. This cultural environment consists of other memes which are also being selected. The meme pool therefore comes to have the attributes of an evolutionarily stable set, which new memes find it hard to invade" (1976, 213–14). Blind faith, as Gottwald suggests, is precisely that, an "alienating line of tradition which absolutizes and falsely projects the traditional religious models into eternal idols and specters of the mind" (1979, 705). The religious term for this misplaced faith is *idolatry*.

Religious belief is always located in a determinate context. Those who dare to confront the reality of their faith also realize that a faith that fails to meet the demands of life, the exigencies of living in a real world, is in danger of collapsing, of becoming empty, irrelevant, inauthentic. Culture is a dynamic phenomenon. Accordingly, demands for innovation are inevitably placed upon religious belief, upon the sacred canopy itself. Here, then, is a supreme irony, for many of the faithful have feared science and philosophy. But an informed grasp of science allows the Great Code to be read intertextually. People of faith need remember that Western culture itself is simply one form among many and, crucially, that it remains in process. We have not arrived at any final sociocultural destination. Americans are both conditioned by and conditioning that history of effects which is culture. Which implies that religion, however irrelevant it might appear to the secular environmentalist, is vital to the creation of a sustainable society.

Judeo-Christianity is unique in its ability to reweave itself in the light of changing circumstances without losing its fundamental inspiration. Sociobiology, so often misunderstood as deterministic, as reducing human freedom to a genetic mechanism, underscores this premise. Wilson argues that a sociobiological account of human beingness does not preclude but in fact explains free will (1983, 182). Whatever the resistance of religious meme-complexes to change, manifest in the orthodoxy of religious practice and institutions, human beings can defy, as Dawkins suggests, "the selfish memes of our indoctrination. We can even discuss ways of deliberately cultivating and nurturing pure, disinterested altruism . . . [and evolving beyond our secular faith that we are *Homo oeconomicus*]. We, alone on earth, can rebel against the tyranny of the selfish replicators" (1976, 215).

Any number of examples from American history alone, from abolition to the civil rights movement, confirm these ideas. The social gospel movement, a liberal movement within American Protestantism, clearly shows how basic Christian doctrine can reweave itself to meet changed circumstances of existence. In this case, Washington Gladden (1836–1918) and Walter Rauschenbusch (1861–1928) applied the biblical tradition to the social problems that grew out of the rapid industrialization of American society. The social gospel movement can be construed as a countervailing force to social Darwinism, which legitimated enormous disparities in income between individuals and enormous disparities in wealth between classes by an appeal to the Darwinian law of competition and survival of the fittest. The social gospel movement perhaps culminated in the New Deal legislation of the 1930s, which embodied in legislation the ideals of the social gospel: Christian love of and concern for the poor.

John Taylor notes that "each time the cultural base" of Christianity has been about to collapse, it "has been saved by its diffusion into a new milieu." If this "infinite translatability" sometimes upsets conservatives, the guardians of orthodoxy, "let them also find reassurance from this history. For it shows that, whatever undreamt-of forms the faith has evolved in each new phase, the dominant themes of former times have been added to rather than lost" (1990, 641). Granted, secular culture sometimes appears to have overwhelmed the Judeo-Christian tradition. The political economy of the late twentieth century is a long way from the Judeo-Christian vision of the founders of American society. The Bible today seems to skeptics as little more than something to be tucked under one's arm on the way to Sunday school. Certainly it contains no solutions per se for pollution buried within. How can anyone claim that the biblical tradition is essential to our future? Surely the experts, not the biblically literate, will rule tomorrow's society.

Perhaps things are not so simple. David Tracy argues that when a literate culture finds itself in a desperate plight the basic question is how to read its fundamental texts. "The once stable author," as he puts it (not altogether facetiously), "has been replaced by the unstable reader" (1987, 12). Furthermore, as George Lindbeck asserts, religious innovation is not only a possibility but a necessity that results "from the interactions of a cultural-linguistic system with changing situations." People modify their faith, he continues, "because a particular religious interpretive scheme (embodied . . . in religious practice and belief) develops anomalies in its application in new contexts" (1984, 39). And, as Geertz suggests, religious faith—grasped in its cultural significance—leads to moral beliefs and practical consequences because it locates "proximate acts in ultimate contexts that makes religion, frequently at least, socially so powerful. It alters, often radically, the whole landscape presented to common sense, alters it in such a way that the moods and motivations induced by religious practice seem themselves supremely practical, the only sensible ones to adopt given the way things 'really' are" (1973, 122).

The point is that religion itself is caught up in the hermeneutic circle. In a time of crisis, literate cultures must fall back on their exemplary texts. Given the cogency of this thesis, any solution to ecocrisis begins with rereading the classics. Which is to say that we don't reinvent the wheel of culture but roll it on from where we are. Regardless of personal biography, our faith, we are caught in a culture where religion has had momentous consequences. Indeed, as secular environmentalists argue, religion is deeply implicated in ecocrisis. More important, it is involved in any solution. Thinking of ourselves as biologically underdetermined and culturally overdetermined makes clear the dilemma we face. On one hand, as we face up to the reality of ecocrisis, we also realize there is no human nature to which we might retreat and find safe haven. Simultaneously, the very modern culture through which we live out our lives is ecologically pathological.

But there is good reason to think that religious believers are capable of rising to the challenge. Judeo-Christianity has always been a critical tradition: it consumes myth at a prodigious rate, since God alone is sacred. Exodus 20.4 makes the point: " 'You shall not make for yourself a graven image, or any likeness of anything that is in heaven above, or that is in the earth and beneath, or that is in the water under the earth.' " Not only is the worship of graven images of either animals or humans forbidden, but as Schneidau argues, "Hebrew institutions, even if divinely ordained, have no inherent sacredness and can always be ultimately questioned. . . . No Hebrew institution can be sacred, because if so it would become an idol" (1976, 4). One merit of the Great Code, as Tracy insists, is that the Bible "has functioned with extraordinary flexibility," lending itself time and again to solutions that maintain tradition even in response to the novel circumstances or exigencies of life (1979, 13). And this, after all, is the fundamental question of religion. As Alfred North Whitehead argues, the religious question is whether "the process of the temporal world passes into the formation of other actualities, bound together in an order in which novelty does not mean loss" (1979, 340). The Bible, in short, is a classic (it endures) precisely because of its surplus of meaning, which engenders the possibility of its renewal. Or, more generally, the Bible is protean; religious discourse is alive, to paraphrase Joel Weinsheimer, forming new concepts as "words are applied to new circumstances in new times" (1991, 118). Chapters 4, 5, and 6 may be read as extending this argument.

Religion and Cultural Evolution

Looking beneath the facade of ecological dysfunction and our collective efforts to practice conservation since Earth Day 1970, a few people have seen something so faint, so vague and ephemeral as to be almost invisible: *the modern world devalues nature*. More explicitly, in modern culture nature has become an economic commodity. Whatever the biblical tradition or other religions might say about the Creation, its value is established by the market—the "clearing" or "space" created by the modern narrative. The modern American is culturally conditioned to think of nature as nothing more than matter-in-motion, as a standing reserve that through technological and entrepreneurial prowess is converted into a consumer's cornucopia. Even so-called nature lovers are often little more than consumers, finding in wilderness refuges, national parks, and mountain ski resorts the "authentic experiences" their lives otherwise lack. Outdoor recreation in America is, after all, big business. Our prevailing worldview rests on the assumption that nature is nothing more than raw material to serve the ends of one species: ours.

Americans are embedded in *a final vocabulary* that offers ready-made descriptions of our relations with nature in a way that closes rather than opens discourse about caring for creation.[8] We have collectively become Homo oeconomicus, that is, actors upon the stage of history. Our preferences as consumers are institutionalized in the marketplace—a shrine built, to paraphrase Marshall Sahlins (1972), to the unattainable goal of infinite needs. Our first language, the discourse that binds secular society, is the language of utilitarian individualism. Insofar as Americans have a common faith, it is the belief that the good life is tied to an ever increasing quantity of life. The high priests of this civil religion are ensconced in the citadels of power: the Federal Reserve System, the Council of Economic Advisors, the Office of Management and Budget, the White House. These experts maintain and periodically refurbish the legitimating narrative of economics—a co-evolved meme-complex—that tells us the Gross National Product and Rate of Economic Growth are holy, unqualified, and absolute societal goods. This secular narrative has, ironically, become a cultural absolute, defining Americans collectively, as a people, regardless of our own faith commitments: the privatization of religion is a fact. Woe be unto the administrator, the politician, and the political party that are perceived as unable to manage the economy and thereby upset economic growth. Government exists above all else to keep the economic structure of society running smoothly. As Alasdair MacIntyre puts it, "The fetishism of commodities has been supplemented by another just as important fetishism, that of bureaucratic skills. . . . The realm of managerial expertise is one in which what purport to be objectively grounded claims [e.g., to the knowledge of the good society and how to achieve it] function in fact as expression of arbitrary, but disguised, will and preference" (1984, 107).

Just as mass and energy are the organizing principles of modern physics, Homo oeconomicus is the theoretical construct around which modern economics orbits and secular culture is built. True, Judeo-Christianity provided a legitimating rationale for the capitalist transformation of Western society.[9] But religion no longer plays a role in the normal operation of the corporate state. Insofar as Americans have a *collective identity* it is as Homo oeconomicus—the mass person, the consumer who lives amid unprecedented material splendor and the producer who bends the earth to virtually unrestrained human purpose. But there are problems with this story. In the first place, Homo oeconomicus is a distorted definition of human beings considered individually and socially. Further, the fact that the market society exists, and that it is so carefully managed by the high priests of economics, does not in any way legitimate it. Again, to quote MacIntyre, a penetrating cultural critic, the truth is that in our modern culture "we know of no organized movement towards power which is not bureaucratic and managerial in mode and we know of no justifications for authority which are not Weberian in form" (1984, 109). Efficiency in achieving a single end, the maximization of the production and consumption cycle, is the criterion used to judge the political and corporate elite.

According to Cobb and Daly, the experts specifically, and American citizens more generally, believe that the common good is "identical with the summation of the increase of goods and services acquired by individual members. The society as such does not appear" (1989, 161). As members of a market society, enabled by the language of utilitarian individualism, we lose sight of any sense of *person-in-community*, any recognition that "the well-being of a community as a whole is constitutive of each person's welfare. . . . These relationships cannot be exchanged in a market" (164). The consequence of our dominating first language is that, as Americans have individually profited from the economy, the social fabric has been increasingly rent. By psychological necessity, affluent Americans are utilitarian individualists, believing above all else in a Smithian calculus (as metaphorically represented by the invisible hand in *The Wealth of Nations*) that objectively and fairly determines their lot in life. Politically, the majority vote their pocketbook, for candidates perceived as representing their best economic interests. One need only consider the presidential campaign of 1992 and the overwhelming concern of Americans with economic issues to confirm this. As Lance deHaven-Smith argues, given the Dominant Social Matrix, there is no reason to expect action on environmental issues. "The average person considers how a particular proposal will affect his [or her] income, lifestyle, or general situation" (95). And the average person also believes, falsely, that ecology and economy are either competing or even antithetical interests. The high priests of economics "cook the books," through the so-called discount rate, to create the opposite impression, while the true costs of not moving toward sustainability exceed the benefits of short-term exploitation.[10]

Many Americans, even those who live below the so-called poverty line, aspire to no greater dreams than those of expressive individualism. *Expressive individualism*, the freedom to create whatever way of living the individual desires, is today's dream of success. Americans tend to define personal success in terms of affluence and its concomitant, the power to create a life of comfort, convenience, and status.[11] The frenzied attempts by Americans to create meaningful lives for themselves perhaps underscores the truth of Thoreau's insight: most people live lives of quiet desperation. The steady growth in the indexes of human misery—alcoholism, drug addiction, crime, depression, and so on—suggest that *la dolce vita* eludes more and more Americans in spite of their affluence.

Whatever the illusions created by the DSM, material success has not been accompanied, as Nicholas Rescher (1974) argues, by an increase in human happiness. A close study of the data, he continues, suggests the reverse of that truth: "postwar progress in matters of human [economic] welfare" has not "been matched by a corresponding advance in human happiness" (95). Staffan Linder (1970) maintains that the struggle for affluence has created a harried leisure class, a nation of materialistic status-seekers so overinvested in the pursuit of wealth that the primary pleasures of life—love, family, community, physical pleasure, and so on—increasingly elude them. Similarly, Michael Argyle (1987) contends that material consumption and income have little to do with human happiness; above the poverty level, variables like the quality of home life, the workplace and its satisfactions, recreation and leisure time, and meaningful friendship and community involvement are far more important than income. The implication is obvious: the quality of life is not tied to its quantity, at least not beyond satisfying basic needs. Whatever the status of economic faith as a secular absolute, and no matter the rituals and incantations of the high priests of controlled capitalism, argues William Blackstone, any realistic solution to environmental crisis entails "a fundamental change in what some call our life-style" (1974, 17).

This will be difficult. Even though the Bible admonishes those who pursue wealth (for example, Eccles. 5.10), many of the faithful are caught up in the chase. Churches themselves often operate market campaigns, pitching "the Gospel" or tailoring the message to the perceived needs of parishioners. The church, the central religious institution of contemporary life, appears to some critics as having been coopted to serve the narrow definitions of individualism.

This is enormously ironic. The culture of individualism appears to be subverting the primary tradition by which it might be transcended. Religion has historically provided an avenue for individuals to involve themselves with community. It now appears that religion is less and less involved with public concerns and more and more simply a matter of one's style of living. Religion has become a commodity, marketed like other products of the consumer society. New churches offer "religious consumers" a "product" that helps them feel good about their lives, success, and affluence.

As the churches have withdrawn from public life, the high priests of economics have replaced value-oriented, qualitative discourse with quantitative considerations of economic growth, per capita income, and so on. The rhetoric of economics has displaced the give and take of open-ended democratic conversation and citizen participation in social affairs. To paraphrase Daniel McCloskey (1985), the rhetoric of economics, modeled on classical science, has become architectonic—a blueprint for modern culture. As the power of economics to set the social agenda has grown, religion has increasingly found itself on the sidelines. Citizen participation in public affairs is vital to the survival of democracy, yet it is arguably at an all-time low. The consequence is a culture at risk because private values—*our choices as consumers*—grounded in utilitarian individualism and expressive individualism, threaten to overwhelm public values—*our choices as citizens*. Tocqueville warned of this danger in the nineteenth century. "Individualism is a calm and considered feeling which disposes each citizen to isolate himself from the mass of his fellows and withdraw into the circle of family and friends; with this little society formed to his taste, he gladly leaves the greater society to look after itself" (quoted in Bellah 1985, 37). Tocqueville was prescient, as were Thoreau and others, for individualism has increasingly deepened the fissure between private and public life.

Yet religion continues to offer Americans a vital and irreplaceable means of social relationship. Further, religious discourse generally, and the myth-ritual-symbol complex of Judeo-Christianity in particular, offers a fundamental means for thinking about society as a whole. The possibility that Americans might fail to use religious discourse in ways that pull them together on social issues is one thing. But the potential for religious discourse to pull Americans together on an environmental agenda is nonetheless real. Actualizing this potential is crucial in a time of crisis. *Crisis* comes from the Indo-European root *krei* (also *ker*), to sift apart (Greek *khrinen*). The root yields many words in English, from *critic* to *curt*. In the context of culture, a time of crisis is the separation of the various elements that constitute the whole. The center will not hold.[12] The context in which the ecological crisis plays itself out is one where Americans feel increasingly isolated, even alienated, from each other and frustrated by the apparent inability of our institutions—primarily the corporation and the government—to do anything about it. The argument that religious discourse has a vital role to play in refashioning society is not itself metaphysical. As I argue in chapter 4, an ethical response to ecocrisis can be based on many differing claims to ultimate knowledge. My aim is to show how people of faith, within the traditions of their own metaphysical commitments, can make vital contributions to public discourse. Solidarity is the issue.

Judeo-Christianity and the Roots of Ecocrisis

The road from there, the deep past of the Paleolithic where humankind lived in harmony with nature, to here, the modern age that gives every sign of having exceeded, as the Club of Rome once stated, "the limits to growth," is long and winding (Meadows 1972). Elsewhere I have presented details of the ecological transition from hunting-gathering society to agriculture and ultimately modern culture.[13] Here a complete exposition is not necessary. What is clear is that hunting-gathering was the predominant form of human existence for some two hundred thousand years or more. This way of life was displaced by herding-farming over a relatively short period of some five to ten thousand years. Carl Friedrich von Weizsäcker observes simply that "civilization meant first of all agriculture. The neolithic revolution changes the forest and steppe into farmland, therefore it tears away the earth from the powers to whom it previously belonged; the peasant has other gods than the hunter" (1988, 75).

By the time of biblical authorship, beginning with Genesis, the Neolithic revolution was history, already some ten thousand years under way. The Old Testament did not lead to the agricultural revolution. Rather it justifies that ecological transition, pulling a sacred canopy over the tribes of Yahweh to legitimate the lives they were living. Set in historical context, Genesis 1.28 comes alive even for the atheist. The consequence of that sociocultural transformation, from our vantage point a process known as the "agricultural revolution," was a radical shift in perspective. Judeo-Christianity, above all else, led us onto the stage of history. Unlike the savage mind, the Judeo-Christian mind found meaning in a time beyond the eternal mythical present.[14] Consequently, the perception of nature as sacred—as the source of value—was abandoned. God the creator was above and separate from nature. Further, God gave dominion to *Man* over the earth.

The Judeo-Christian outlook on the natural world proved potent. Clarence Glacken summarizes this outlook, largely finished by A.D. 300, in *Traces on the Rhodian Shore*. "The Judeo-Christian conceptions of God and of the order of nature . . . [created] a conception of the habitable world of such force, persuasiveness, and resiliency that it could endure as an acceptable interpretation of life, nature, and the earth to the vast majority of the peoples of the Western world until the sixth decade of the nineteenth century" (1967, 168). Which is to say that the Judeo-Christian worldview—or sacred canopy—has shaped the basic outlook of Western culture on nature for nearly two thousand years. No one (atheist or secularist, social scientist or scientist) can truly understand the circumstances of our existence apart from the Great Code. Indeed, apart from realizing that *history of effects*, there is little chance that religion can help us work our way through a time of ecological crisis. What we need to bear in mind, however, is that we are both affected by and affecting history.

As Glacken suggests, the 1860s marked the beginnings of an epochal transformation in Western culture, a process that influenced both the comprehension of our relation to nature and our concept of religion. When we stand back from the so-called environmental crisis, from the immediacy of polluted water and the global greenhouse, and juxtapose it with the Judeo-Christian faith that was fundamental to the creation of the modern world, we discover the scientific untenability of one of its central premises; the reality of ecological crisis has rendered the belief in providence problematic. The scientific judgment that providence is untenable is a stinging blow for people of faith, for it denies humans a place in a designed cosmos, in a universe that is hospitable and friendly toward us rather than malevolent and threatening. Further, the looming possibility of ecocatastrophe undercuts faith in the notion that God designed the earth for Man. The problem is acute for the religious conservative, who believes that Genesis is infallible and therefore irreconcilable with Darwin, and difficult even for the religious liberal, who believes that Genesis and Darwin can be reconciled. Ecocrisis apparently confronts Judeo-Christians with a dilemma: either God's design is flawed or the perception of God's design on the part of his children is flawed. Yet this is to say no more than where science is assumed as a definitive perspective, as one that exhausts cognitive meaning, then the Judeo-Christian notion of providence is necessarily untenable.

My argument does not assume that scientific discourse is a final vocabulary intrinsically superior to religious discourse. In chapter 2, I argued that a realistic approach to ecocrisis requires setting scientific and religious discourse together in collaboration. Chapter 4 offers evidence that this can be done. In this chapter I am claiming that the Great Code offers opportunity for a new metaphor—*caring for creation*—and that this metaphor is the toehold necessary for solidarity in a time of ecocrisis. Granted, as some of my readers have pointed out, such a description of religious discourse is problematic for religious conservatives and even moderates, since they are vested in claims to ultimate knowledge. From their perspective, religion offers sure and certain conviction in knowing the nature of things. But such claims to ultimate knowledge in no way preclude the faithful from participating in public discourse that speaks to ecocrisis. I claim that across the spectrum of faith, almost without exception, the faithful are given ample reason to care for the Creation and that through that caring we can come together on an environmental agenda. Religious discourse offers the best chance that ecology might influence public policy. By assuming its role in a time of ecocrisis, religion might reestablish its place in public discourse, its vital place in the maintenance of our world. Indeed, from my perspective, religious discourse necessarily has a place.

Intellectuals, including ecologists, ecophilosophers, economists, and experts of all hues and stripes, like to think that maintaining culture depends on the acts of their intellects. But if, as Berger puts it, "legitimation always had to consist of theoretically coherent propositions, it would support the social order only for that minority of intellectuals that have such theoretical interests—obviously not a very practical program. Most legitimation, consequently, is pretheoretical in character" (1967, 30). Institutions, for example, are legitimate simply because they exist. The specious present leads individuals to believe that the existing institutional structure of a society has always been and will always be there. But religious discourse provides the ultimate legitimation for institutions "by bestowing upon them an ultimately valid ontological status, that is, by *locating* them within a sacred and cosmic frame of reference" (Berger 1967, 33).

Religion as Public Discourse: On the Road to Solidarity

Public policy is presently conditioned primarily by our choices as consumers. Utilitarian individualism rules, and so the individual pursuit of economic success overrides all other considerations. Whereas countervailing forms of discourse once existed, today the hegemony of economics is unchallenged. The government can hardly be described as restraining the ecologically destructive consequences of economic growth, for the actions of the state are guided by the same language that steers the private sector.[15] This is not surprising, since capitalism is founded on Adam Smith's *Wealth of Nations*, and it in turn is based on a Cartesian-Newtonian paradigm. As Ernst Mayr, the noted historian of biological thought, observes, this "was a tragedy both for biology and for mankind" (1982, 79). Our basic institutional arrangements (political economy) are founded on the principles of classical physics. Human beings, so viewed, are no more than selfish atoms pursuing private interests.

In the twentieth century we have witnessed the emergence of an increasingly rationalized, planned society controlled by experts. *This elite cadre now rules our society with the legitimating narratives of economic discourse.*[16] As the Great Society and unprecedented wealth have appeared, the language of utilitarian individualism has increasingly pushed religious discourse to the margins. In the flush of material success, engendered by "expert cognition" and "scientific management," Americans have lost any sense of an older tradition of self-governance—the rhetorical tradition. This tradition empowered democracy by bringing citizens together in conversation about the common good. Today the cultural conversation is in disrepair. Americans live in a managed society in which experts determine the means to one pre-established institutionalized end: the maximization of wealth.[17] But something vital to democratic life is missing: discourse that considers issues of the common good beyond consumer preferences. With the rise of the expert and the advent of the managed society, such conversation has been either displaced (marginalized) or replaced (excluded from the policy-making process). Scientific cognition (linear, reductionistic, and quantitative), itself a complement of utilitarian individualism, has overridden the rhetorical tradition, now viewed as idealistic, qualitative, and ineffectual. To its proponents, the political and economic managerial elite, economic discourse seems objective and impartial (value free), whereas religious discourse is subjective and partisan (metaphysical).

The problem is not with scientific cognition per se but with the reification of its "methods" into an absolute. As Davis and Hersh argue, there is nothing intrinsically wrong with mathematizing physical and social phenomena. The difficulty is that "too much of it may not be good for us" (1986, xv). Or, as Irving Kaplan (1964) notes, the danger in mathematics is the law of the instrument. Give a small boy a hammer, and everything he sees needs to be hammered. Or, finally, as Georgescu-Roegen argues, the modern mind is haunted by the arithmomorphic fallacy—the belief that the only meaningful concepts are ones capable of mathematical elucidation. "An economic model, being only a simile, can be a guide only for the initiated who has acquired an analytic insight through some laborious training" (1971, 333). Most experts fail to realize this, believing the economic model to be "an accurate blueprint," "an objective model of reality."

Economic models and expert discourse are just that: no more and no less. *By recognizing the role of religion in a time of ecological crisis, the democratic citizenry can ensure that a genuine debate takes place—a truly democratic, open-ended conversation that brings the fundamental questions of the good society back into public discussion.* Every tradition of faith can find the grounds within itself to care for creation—a caring that is the beginning of a solution for ecocrisis. By caring for creation we can begin to reestablish ourselves as a moral community, as a nation with some collective sense of who we are and where we are going, even though this community grows out of diverse traditions of faith. Public discourse about caring for creation is precisely that: public and not metaphysical. What ultimately counts is a conversation about public policy that is something more than expert discourse. Most crucial, apropos of my thesis, is nurturing the kind of discourse that is inspired by religious narrative rather than market preferences, especially those narratives that focus on caring for creation. Any movement toward sustainability requires a political consensus, the will to act, that does not now exist. The prevailing definition of the public good is stuck inside the conversation of utilitarian individualism. Religious narrative is our best chance to break free of this as a democratic people.

"Caring for Creation" as Metaphor

The phrase "caring for creation" should be understood as a powerful metaphorical and therefore protean expression. Metaphor, under the onslaught of positivism, has been denigrated. Philosophers, charging that metaphors are semantically confused, and scientists, claiming that the only meaningful concepts are those that can be expressed quantitatively, have been especially hostile toward metaphor. But as Northrop Frye points out, "It seems clear that the descriptive phase [of language] also has limitations, in a world where its distinction of subject and object so often does not work. There is no question of giving up descriptive language, only of relating it to a broader spectrum of verbal expression" (1981, 17).

Today the function of metaphor is being reassessed, inspired in part by the work of scholars interested in rhetoric, including Kenneth Burke (1966) in literary criticism, Richard McKeon (1987) in philosophy, and Julia Kristeva ([1981] 1989) in linguistics. In the last two decades numerous inquiries into the role of metaphor in language and knowledge more generally, including scientific knowledge, have been made by philosophers including Paul Ricoeur

([1975] 1977) and Donald Davidson (1984). Mary Hesse (1980) argues, for example, that without metaphor, scientific knowledge is an impossibility, for inquiry would effectively be frozen into place. With the contemporary reawakening of interest in metaphor has come the attempt to recover the ancient teachings, including those of Aristotle, perhaps the first thinker to consider metaphor seriously. "The greatest thing by far," as he put it, "is to be a master of metaphor. It is the one thing that cannot be learnt from others; and it is also a sign of genius, since a good metaphor implies an intuitive perception of the similarity in dissimilars" (*Poetics*, 1459a). Implicit in Aristotle's observation is some germinal notion of the potency of the metaphor of caring for creation.

Twentieth-century thinkers take us beyond Aristotle. George Lakoff and Mark Johnson assert that "metaphor is pervasive in everyday life, not just in language but in thought and action" (1980, 3). What they grasp and Aristotle perhaps misses is that metaphor has a creative dimension that extends language beyond the rhetorical fashions of an age. Metaphor, claims Colin Turbayne, "is the only way we have for saying something new" ([1962] 1971, xii). Metaphor, as Davidson puts it, "is the dreamwork of language and, like all dreamwork, its interpretation reflects as much on the interpreter as on the originator. . . . Understanding a metaphor is as much a creative endeavor as making a metaphor, and as little guided by rules" (1984, 245). Post-positivism, metaphor has reemerged as central to the operation of language, especially to the cutting edge of human self-knowledge. Lakoff and Johnson argue that the most basic conceptual systems that structure the way we think and act are "fundamentally metaphorical in nature" (1980, 3).

The study of metaphor by semioticians exceeds the scope of my inquiry.[18] But, among other functions, metaphor is a means whereby (1) change in meaning is created and (2) the perception of similarity amid difference is identified. Nietzsche caught part of this idea in his notion that truth is a mobile army of metaphors.[19] Wittgenstein, one of this century's most insightful theorists of language, argues that language games—conceptual systems like physics or Judeo-Christianity—are actually forms of life: responses to the conditions of existence. But language, as Heidegger insists, both reveals and conceals. Such a view does not "divinize" language. It does question "representationalism"—the idea that language is a transparent medium that opens a window onto the real world of timeless truth. As Rorty suggests in *Philosophy and the Mirror of Nature*, conversation is the context within which knowledge can be understood. So viewed, for human beings there is no god's eye view of the world, the things in the world, or the relations among the things in the world. It follows, insofar as human beings are to adapt to changing circumstances of existence, that language must itself evolve. Metaphor is one of the crucial variables if not the primary factor in linguistic evolution. Wittgenstein suggests that "a new word [or metaphor] is like a fresh seed sewn on the ground of the discussion" (1980, 2e).

Perhaps, in the beginning, knowledge is metaphoric. Metaphors are among *the causes* of our ability to know more about and respond to the world—by enabling either new theories or legitimating narratives—rather than expressions of such knowledge. Rorty distinguishes between "the realm of meaning," that is, a "cleared area" or established set of conventions within "the jungle of use," and that which falls outside the established boundaries of the domain of meaning: namely, metaphors (1991b, 164). Metaphors in effect refer to "uncleared space" because, though they are alive they are unparaphrasable; "they fall outside the cleared area." So viewed, metaphors are like surprises. "We come to understand metaphors in the same way that we come to understand anomalous natural phenomena. We do so by revising our theories so as to fit them around new material. We interpret metaphors in the same sense in which we interpret such anomalies—by casting around for possible revision in our theories which may help to handle the surprises" (167).

The metaphor of caring for creation can be, for the present generation, a new word sown on the ground of discussion, an inroad to a new form of life, the beginning of a new story that acknowledges nature as something more than resource for economic growth. *More specifically, I am claiming that the biblical tradition specifically and religion more generally has an overabundance of meaning that can yield—indeed, is already yielding—new interpretations of our relations to the earth built around the metaphor of caring for creation.* Religious discourse can help Americans recognize a common good, regardless of the diversity of our claims to ultimate knowledge, that legitimates redirection of our society toward sustainability. The great power of the metaphoric expression "caring for creation" helps establish a public value, a social preference.

This metaphor does not entail any attempt to engender a master narrative, such as a theology of nature. Caring for creation might lead to a variety of either theologies of nature or natural theologies, and indeed, there are some that already systematically express such concern. George Hendry's *Theology of Nature* is one, written at least partly in response to the charges that Lynn White leveled against Judeo-Christianity. According to Hendry, the question for theology "is the place, meaning, and purpose of the world of nature in the overall plan of God in creation and redemption?" (1980, 11). Clearly, a theology of nature might systematically unpack the metaphor of caring for creation. But this step is not immediately relevant to building solidarity. Indeed, the paradox of environmentalism confronts not only systematic environmental ethics but existing theologies of nature. As mentioned earlier, social legitimation is, whatever the opinion of intellectuals, more often than not pretheoretical. I do not contest the fact that a metaphor of caring for creation can, and perhaps should, be extended to a theology of nature. But it can empower people to act independently of a theology of nature. It might, for example, lead responsible Judeo-Christians to address issues, like pollutants in the workplace and in the home, in an ethical manner on the basis of established biblical models.[20]

Neither does my use of the metaphor of caring for creation require an argument that it is the master metaphor, the root metaphor, of the Bible. It could lead to such a discussion (and Hendry explores this idea to some extent), but I do not claim here that it is the master metaphor of the Bible. The goal is to enable religious believers, across the Judeo-Christian spectrum and beyond, to discuss their obligations vis-à-vis the natural world. Certainly a case could be made that the Creation is a root metaphor, a comprehensive conceptual system that embraces all that was, is, or ever will be, including human personality. But caring for creation does not require this kind of conceptual exercise. Rather, it entails only that the faithful look to their faith with new eyes and find therein the words that speak of loving relations with and our responsibilities for the Creation.

Although the metaphor of caring for creation can be interpreted in many ways—that is, it means different things within distinctive faith traditions—at a political level it implies that in spite of our religious diversity we can achieve solidarity on environmental issues. Caring for creation, in short, portends green politics. Social regulation is more than just laws or rules or policies. It is better grasped, as already discussed, as expressing what we believe and stand for as a people rather than as individuals. As Sagoff puts it, "Social regulation reflects public values we choose collectively, and these may conflict with wants and interests we pursue individually" (1988, 17). Neither methods of cost-benefit analysis nor the market mechanism itself can substitute for deliberation, for the rhetorical tradition that allows diverse members of the community to speak together in the quest for comprehension of the common good. Utopian capitalism, as Sagoff makes clear, is moribund. Not all preferences in life nor all measures of value are expressed through or captured by the market. Or, argues Sagoff, "we are not forced to accept the vision of society associated with contemporary economic theory; we are not required to adopt that conception of morality and rationality in preference to some other list of moral, cultural, or historical values. This is not merely to say that we can, in an act of will, replace one morality [utilitarian individualism] with another. It is to insist, rather, that we can see things differently, that we can see the good not as an object of preference but as an object of shared belief, insight, and affection" (122).

In other words, the goals of environmentalism do not derive from or express the self-interest of consumers, that is, a mere willingness to pay for clean air or water or wild and open spaces. *The standard economic interpretation of environmentalism deceives us.* Such goals as the preservation of biodiversity and the reduction of atmospheric CO_2 represent the aspirations of Americans as citizens rather than their willingness to pay for a commodity or service. They also reflect a concern for goods beyond narrowly economic self-interest, such as social justice for future generations. Perhaps, too, an environmental agenda reflects the American experience, the memory of almost five hundred years of constant give and take with nature. Whatever else our specifically American attitudes toward the earth imply, they in turn reflect a much older and more abiding Judeo-Christian history of effects. As I have suggested, Lynn White's analysis of the Judeo-Christian roots of environmental crisis is flawed in some ways. But he correctly maintains that we cannot comprehend ecological malaise apart from this narrative tradition. And I am arguing that Americans cannot extricate themselves from ecocrisis without engaging this history of effects. At least some environmental ethicists, such as Eugene Hargrove, agree. He maintains (1992) that the best way to define environmental values is to tie them to our cultural heritage—to see, for example, caring for creation as derived from tradition.

So viewed, the question is this: Can the church help the American polity develop values that would make environmentalism a political force? Time will tell. But there is no reason in principle that religious discourse cannot make the crucial difference. Almost without exception, people of faith observe that consumptive ways of life are a primary cause of ecocrisis. The elite managers of modern society reflect this social reality: *the state panders to our preferences as consumers.* But we have an alternative. The biblical tradition offers Judeo-Christians ample resources to find reason to care for creation. Through the Great Code, particularly through the metaphor of caring for creation, we might reweave the scared canopy. Such an approach does not either marginalize any faith tradition or require that believers disavow claims to ultimate knowledge. By developing that metaphor in religiously distinctive ways we can begin collectively to refashion our relations to the earth. And, as we shall see, this metaphor can without exception be enriched by science.

Recontextualizing Religion and Environmentalism

Religion per se no longer exercises a significant influence in American politics: a so-called *civil religion* dominates. The Dominant Social Matrix (discussed in chapter 2; see table 1) sketches the broad outlines of our civil religion. Proponents of the civil religion believe that *economic progress is a law of nature*, that technological innovation will ultimately resolve any sociocultural problem and that America (whatever its flaws) epitomizes the perfect society. Insofar as the civil religion is dominant, then the potential for political response to the environmental crisis is small, since voters—of whatever religious affiliation—will "vote their pocketbooks," that is, their own short-term economic concerns, even if these are societally self-defeating in the long run.

The American civil religion goes hand in hand with utilitarian individualism. The DSM is a secular article of faith, masked by the rhetoric of objectivity: a final vocabulary insulated from an open-ended *cultural conversation* that addresses public issues beyond those of producers and consumers. Thus, for example, the belief that progress is a law of nature is accepted uncritically, independent of arguments and evidence that count either for or against it. Criticism of the dream of infinite economic progress, such as Nicholas Georgescu-Roegen's *Entropy Law and the Economic Process*, largely goes unmet. How, he asks, can unlimited economic growth be sustained in light of the reality of the second law of thermodynamics? How can economic growth, fueled by the ever-increasing consumption of fossil fuels, be reconciled with the global greenhouse? How can biodiversity be protected when the human population continues to grow exponentially? The answer from the high priests of the American civil religion—the economists who staff the Council of Economic Advisors, tend the National Income Accounts, and more generally attempt to manage the economy—has been the deafening sound of one hand clapping.[21] Those few proponents of the DSM who have entertained such questions, such as Julian Simon and Herman Kahn, reiterate their faith that technological solutions exist for all limits to growth.

Environmentalism has had some influence on the political process: the National Environmental Policy Act, the Endangered Species Act, and the Alaska Wilderness Act are a few signs. Yet, as Lance deHaven-Smith suggests, there is good reason to think that environmentalism will be "transformed into an institutionalized movement that is structurally and ideologically incorporated into existing social and political relations" (1991, 4). Thus, the fundamental issues posed by ecocrisis go begging—questions concerning national energy policy, optimum population size, the global greenhouse, acid rain, and so on. What "began as a radical challenge to the capitalist system," continues deHaven-Smith, "rejecting consumption, unrestricted economic growth, and exploitation of the natural environment," is politically and economically coopted "by making marginal adjustments to economic activity" (4). The paradox of environmentalism is partly explained by this reality. Our liberal-democratic society careens onward despite scientific evidence that the present path leads to ecocatastrophe. We also appear to be a nation headed toward social catastrophe, what Tocqueville calls *administrative despotism*, an "orderly, gentle, peaceful slavery . . . established even under the shadow of the sovereignty of the people" (quoted in Bellah 1985, 209). The signs are many; witness the growth of national planning (including the politicians who create and the experts who implement such schemes), the increasing cooperation between the corporation and the state (euphemistically called the "new industrial state," the "corporate state," or "controlled capitalism"), and the growing skepticism of politics and consequent withdrawal from public life by citizens.

Ironically, the avoidance of administrative despotism does not involve abolition of federal and other levels of government. The so-called invisible complexity of postindustrial society likely precludes that strategy.[22] Instead, two things must be done, both of them consistent with environmentalism. First, we must strengthen the organizations through which ordinary citizens can influence the processes that establish the social agenda. Michael McCann argues that one of the weaknesses of environmentalism has been its lack of awareness of the importance of organizational forms as influencing "effective citizen mobilization." Public interest groups like the Sierra Club, the Nature Conservancy, and the Wilderness Society have achieved large memberships; but the vast majority of their members remain uninvolved beyond sending in an annual check. The key factor that has been overlooked, according to McCann, is that "structural relations within groups [are] a major factor in shaping citizen perceptions of options, incentives, and purposes at stake in the issues" (1986, 175). My claim, which extends McCann's argument, is that the church, which not only has the weight of tradition but also provides a sacred canopy for its members, is the most likely organization for creating solidarity on sustainability.

For one reason, *the church is perhaps the only institution in modern society capable of resisting administrative despotism* (see chapter 5, below). That is, the church empowers citizens with the language (a moral tradition, a sacred canopy, a sense of ultimate knowledge) that enables them to question the rule of our collective lives by a managerial elite pursuing economic expansion for the sake of growth alone. The strengthening of the church is fully consistent with liberal democracy, or polyarchy, as Dahl terms it. For another reason, again following McCann's lead, "human choices cannot be understood apart from the meanings, values, and expectations that evolve out of collective life" (1986, 175). The church offers precisely such a context, and hence the potential of expanding politics beyond mere procedures to issues of ethical substance. Since World War I, the American government has relentlessly rationalized the processes of governance: we live in an era of scientific management. What has been lost in this rationalization is any deliberation over the ends of society, the collective goals that weld a diverse people into a culture. But within the biblical tradition such questions might be entertained by ordinary citizens. As noted previously, those individuals who serve the state do not deliberate over the ends they serve, and perhaps rightly so: their function is to manage rather than determine the social agenda.

John Cobb argues that these two actions (revitalizing public interest institutions, including the church, and a politics of substance) are not necessarily independent. The church, he maintains, has the capacity to deal with social issues, including those concerning the environment. And in dealing with these issues it can influence "other nongovernmental organizations with which it needs to work closely. Rightly directing the energies of these private institutions may be as important as directly influencing government policy. Often government policy will follow directions pioneered by other institutions" (1979, 153). It follows that environmentalists should consider the possibility if not the probability that religion has a vital role to play in setting an agenda for the environment. The alternative is a totally "rationalized" and "managed" society.[23] As experts insist, environmental issues do involve technical expertise—in fact finding, in devising technologies to protect the earth, and in policy implementation. But what might be gained by the movement to care for creation is a refocusing of environmentalist discourse from technical and market questions to value questions, questions about who we are and where we are going as a people.

Many reasons are offered as to why the ecological problems confronting America remain unresolved. One is a lack of consensus and any sense of purpose among voters about what needs to be done. Environmentalists, as Lance deHaven-Smith points out, tend to believe that their interests reflect "a stable and widely shared orientation of modern mass publics." In truth, he claims, environmentalism "is a movement whose philosophy is only skin deep. Beneath the scientific arguments and ecological warnings of environmental interest groups are voters with very narrow and mundane concerns" (1991, 93). Primarily, their concerns are narrowly economic, which means that the American civil religion dominates public policy. Most people consider environmental proposals only in economic terms. McCann amplifies deHaven-Smith's point: environmentalism has had neither the organizational structure nor the moral authority to challenge the corporate state. "Shunning the programmatic concern for such basic exigencies of economic policy and majoritarian political organization, therefore, the public interest [environmental] reform agenda has remained a mostly peripheral force in American public life" (McCann 1986, 25). Creating solidarity on sustainability cannot be either realistically envisioned or empirically described apart from an organizational framework in which citizens actually begin to grapple with environmental issues in terms of an ethic that is autonomous from the American civil religion.

The work of Mancur Olson poses a further challenge to environmentalism, since he argues that there are reasons to think, even granted the premise that individuals make rational decisions, that groups composed of such individuals will not make rational decisions. "Indeed, unless the number of individuals in a group is quite small, or unless there is coercion or some other special device to make individuals act in their common interest, *rational, self-interested individuals will not act to achieve their common or group interests*" (1965, 1–2). The tragedy of the commons, that is, the unrestrained exploitation of goods not owned by anyone, such as the air and the water, reflects this reality. In the context of utopian capitalism it is never rational for a self-interested person or business to sacrifice for a collective good. Environmentally aware manufacturers who voluntarily incur additional production costs by meeting objectives like pollution control, for example, will be driven from the market by other manufacturers who exploit the commons, charge lower prices for their goods, and let consumers make their own choices. On the flip side of the coin, it is rational for consumers to purchase lower-priced goods (plywood from the Amazon basin rather than from the Pacific Northwest) or services regardless of environmental consequences or repercussions for future generations. Consumers pay a premium for "environmentally sensitive" products.

As McCann notes, on Olson's analysis "the larger the group that benefits from a collective good, the less is the individual's fractional share of received benefit for organizational input, and the less is the incentive to participate for collective goods rational in its essence" (1986, 174). This helps to explain why so many Americans appear indifferent to "the environmental commons," clean air and water, biodiversity, and so on: however rational a goal the preservation of the earth is, it remains a collective good. Ironically, Olson's argument underscores the importance of religion. Olson assumes that human beings are motivated primarily by economic gain; but this is not true. Religion offers a language that remains outside economic discourse. And it is a language most Americans speak.

Yet religion has been pushed to the edge of the political process. Johnston points out that elections in liberal-democratic society are conducted "in the context of a dominant ideology created and upheld by the state into which people are socialised, and around salient aspects [such as the accumulation of wealth] of which . . . they are mobilised by the major political parties" (1989, 165). Given this reality (co-adapted meme-complex), there is little reason to think that environmentalism—as traditionally conceived—can have much effect. But insofar as religion brackets the American civil religion, it will play a decisive role in resolving the ecological crisis. Religious discourse, in short, is potentially more important than secular environmentalism, at least for the immediate future, because of the influence that people of faith might have on the political process. In empirical terms, a social consensus on the importance of caring for creation would be reflected by candidates (in their platforms and actions), since a failure to do so would alienate a majority of voters or at least a solid voting block. Alternatively stated, the greening of religion portends changes in public policy that secular environmentalists have been unable to achieve, whatever their philosophical, ecological, and economic arguments.

Environmentalism may converge around a rejection of Homo oeconomicus. Environmentalists, Paehlke argues, "prefer social, psychological, spiritual, and symbolic fulfillment of human needs to fulfillment via commodities" (1988, 206). Religiously inspired discourse might include discussion of the social, moral, psychological, and spiritual implications of environmentalism. John Firor argues that if we are to avoid administrative despotism and turn away from the high-impact, technological path that seeks to dominate earth, then we must somehow develop "an alternative definition of what it means to be human on earth" (1990, 125). George Brockway (1985) contends that Americans are beginning to reject the notion that things—that is, material possessions, commodities, the GNP, and the rate of economic growth—are more important than people—that is, the quality of life, the physical, intellectual, and spiritual well-being of humans. Similarly, Rescher (1974) argues (as noted earlier) that once beyond satisfaction of primary needs such as food, housing, health, and education, there is little correlation between higher standards of living and happiness. Granted, it is difficult to specify how much housing or health care or education meets basic needs. But that is not the issue, since most middle-class Americans enjoy adequate levels and all Americans could be afforded similar opportunities. The issue is rather that the single-minded pursuit of economic growth becomes self-defeating when its only aim is to increase the GNP. The unprecedented post–World War II economic boom has not led to a golden age: Americans report a decline rather than an increase in happiness.

Some environmentalists emphasize the need for leaders, experts who could focus diverse publics on environmental policy. Environmentalism, as Lance deHaven-Smith sees it, "is not a set of [pre-established] attitudes; it is a political process bringing attitudes into a conceptual framework" (1991, 96). He contends that there is no already existing, generalized concern for the environment among American voters. Samuel Hays has a similar position. He contends that environmentalism was "forged not so much by general theory and preconceived thought as by day-to-day concern and action. [The movement] . . . arose from the varied ways people confronted their surroundings and found them either helpful or harmful for the realization of their aspirations" (1987, 246). So viewed, the "progress" on environmental issues claimed by nongovernmental organizations (NGOs) and so-called environmental presidents, such as the Clean Air Act (1990), are more gestures addressing the worst and most visible consequences of economic growth and development than responses to deeper issues.

The emphasis on the local origin of environmental consciousness is important. No leader can lead a democratic people where it does not want to go. The future of a democratic society ultimately falls on the citizens, ordinary folk. If there is going to be change, the citizens almost necessarily run ahead of their leaders and parties. (Singular leaders, like Abraham Lincoln or Martin Luther King, Jr., are able to take the pulse of a nation—on slavery, on civil rights—and lead the people in the direction they are collectively capable of going.) Environmental elites—philosophical, ecological, and political—have assumed that the general public will respond to abstract arguments about, for example, the importance of thinking in terms of ecosystems. But deHaven-Smith counsels a new strategy. "Rather than formulating concepts from the top down, that is, from complex ecological theories down to problems in everyday life, a better strategy is to work from the bottom up. Very mundane concerns must be linked to more general but still rather localized issues before they can be seen to be part of even more complex and distal processes" (1991, 98). I shall show in chapter 4 how people of faith can find good reason within their own religious traditions to care for creation and in chapter 5 how the local church is where the rhetorical tradition might revitalize itself.

The Sacred Canopy and the Good Society

Because we live in a secular society, the economist has replaced the preacher in the social scheme. According to Rorty, "Worries about 'cognitive status' and 'objectivity' are characteristic of a secularized culture in which the scientist replaces the priest" (1991b, 35). In American society the high priest has been the economist and the holy grail has been the GNP. Americans have placed their secular faith in the National Income Accounts, believing them to be a measure of both social progress and individual well-being. Modern economics has become, Daniel McCloskey asserts, a religion with its own Ten Commandments (1985, 7–8). This faith, as Rorty implies, exemplifies the consuming desire for rationality in a strong sense, "a sense which is associated with objective truth, correspondence to reality, and method, and criteria" (37).

A great irony is that the present generation of Americans, the managerial elite and the voting public, insofar as they even entertain questions of environmental policy, reduce them to questions of economics. *But environmental questions are not primarily economic questions: they are first ethical and then political.* For example, the trading of so-called pollution permits on the market, ballyhooed as the way to have clean air, is a "shell game without a pea," since the basic problem (that is, the adverse impact of manufacture upon air quality) is not resolved but merely moved—away from a developed and consequently polluted area to an undeveloped and not yet polluted area.[24] This is "environmentalism"? The earth, in this scenario, remains nothing more than a sink for pollution. And the responsibility of the present generation to future generations is shirked. The trading of pollution permits on the market represents economic expediency rather than a solution.

Though experts may pretend that they can "manage planet Earth," political decisions about environmental policy cannot be reduced to questions of economics and technology. Still, economists like to argue that problems like pollution and the preservation of biodiversity can be solved through the market. It is puzzling that these economists readily admit that such problems are themselves the consequence of economic decisions. Economists like to argue that economic theory describes an objective reality that is out there, independent of human value judgments. Economics, however, is not value free, whatever the pretense of economists. And a market society is a socially constructed reality, merely one among an array of forms of organization. It follows that every dollar of economic growth is not the unqualified good the religion of economics proclaims it to be. Indeed, some forms of economic growth fundamentally undercut the possibility of a good society. It may also be that some *total* level of economic production is counterproductive, that is, an increment of income that diminishes rather than enhances the quality of life. For example, under a strictly economic calculus of value, the unrestrained use of fossil fuels is a social good, since it increases the rate of economic growth. But how can a society that fails to curb its CO_2 emissions claim to be a good society, since the long-term consequences of such behavior effectively cheat future generations of the possibility of a good life? Can any society that manifests a cavalier indifference toward future generations conceivably be good?

Religious discourse offers an opportunity to expand the cultural conversation about environmental issues beyond economics. It offers the possibility to forge a consensus on public policy that overrides the standard economic interpretation of environmentalism. The "willingness to pay" model of environmentalism undercuts the possibility that society might achieve sustainability, since intrinsically ethical issues like air quality, biodiversity, and ecosystem protection are reduced to quantitative measure, secondary to the ultimate social good, the GNP. As we shall discover in chapter 4, paradigms of *imagination* are sometimes more important than paradigms of *inference* in the societal decision-making process. Rorty argues that we should abandon the attempt to divide sentences into one set of scientific statements that have "objective truth" and therefore represent "genuine knowledge," such as economics, and another set of value statements that are "subjective" and therefore "mere opinions," such as religion. So, too, we should abandon any idea that scientists have "special methods" that citizens can employ in making value judgments so that they might have "the same kind of self-confidence about moral ends as we now have about technological means" (1991b, 37).

In Rorty's view, a healthy society is one in which the citizens "replace the desire for objectivity—the desire to be in touch with a reality which is more than some community with which we identify ourselves—with the desire for solidarity with that community" (1991b, 39). Apropos of ecocrisis this means that achieving a consensus on sustainability, that is, solidarity, is more important than economic considerations involving only self-interest, that is, objectivity as defined by the market. As Rorty argues, "Social policy needs no more authority than successful accommodation among individuals [or groups], individuals [or groups] who find themselves heir to the same historical traditions and faced with the same problems" (184). Rorty's critique of representationalism, as I read it, does not deny the importance of objectivity in conversation, that is, issue-specific dialogues governed by the use of evidence and logical criteria. The point is not to sacrifice objectivity on the altar of solidarity but to stress the value-laden nature of the interpretation of the evidence that bears on determining an environmental agenda (see chapter 1 on pollution).

Rorty's argument may strike religionists who make claims to ultimate knowledge as too relativistic, too subjective. But his position neither precludes belief among the faithful in the God-who-is-there nor renders their knowledge claims illicit. All religious believers, including those whose claims to ultimate knowledge are privileged (for example, the claim that the sentence "Our Father who art in Heaven" corresponds with reality), can nevertheless address environmental issues in ways that promote solidarity. That is, in a public or democratic context, as distinct from a private or denominational setting, the toeholds a religion affords are more important than the skyhooks that religion affords to the faithful (though belief in the reality of the sacred is of overriding importance to them). Believers who fail to consider the implications of their faith to care for creation, and to explore the possibilities for solidarity on this issue with other citizens, thereby abdicate their social responsibility to the high priests of economics.

So construed, social progress consists not in comparing society to some ahistorical reality outside the cultural stream but in the adaptive significance of actions in the present. Democratic conversation—the rhetorical tradition—is essential to making social progress, since a consensus to alter the direction of society (to adopt new policies) cannot be otherwise achieved. Without democratic debate leading to solidarity on an environmental agenda, the status quo simply perpetuates itself. Abraham Kaplan argues that policy deliberation in America is undercut by the utilitarian fallacy. "Ethics [value judgment] is thought to bear on policy, not from within, but only at the edges. Our conception of political morality is legalistic: we usually suppose morality to be threatened only when the law is violated. For the mass of citizenry, policy raises moral issues only when its adoption or administration involves bribery, corruption, or venality" (1963, 40).

Described in sociolinguistic terms, no appeal to any reality outside the ongoing process of democratic life is necessary to justify social policy. Which is precisely the reason that the ecologically disastrous and self-defeating pursuit of economic growth for the sake of economic growth alone continues. There are no guarantees that cultural beliefs, however sanctified by tradition and deeply institutionalized, are adaptive, that they have survival value. The environmental crisis is "reality testing"; it is a measure of the ecological feasibility of present social policy. Environmentalists who have repeatedly raised warning flags that we are headed down the primrose path to disaster have failed the American public insofar as they have focused on economic growth in general as the cause of ecocrisis.[25] The fact is that some kinds of economic activities need to grow, such as the solar energy industry and mass transit, and other kinds need to be curtailed, such as the consumption of hydrocarbons. A new generation of ecological economists (as distinct from resource economists) is beginning to develop an economically robust and politically viable theory of sustainability.[26] But the environmentalist critique of the corporate state, whatever its insufficiencies, has served us well by questioning our civil religion.

The rhetorical tradition can be understood as a social practice that optimizes the possibility of new directions and policies in response to the challenges that environmental crisis poses. Religion is singularly suited to help our enormously diverse society forge a consensus on sustainability, to redirect the economy in ways that affirm rather than threaten life. The biblical tradition offers a second language and an organizational structure (a community of memory) that can challenge our first language of utilitarian individualism and the corporate state. Every tradition of faith, including those outside the biblical tradition, can find reason to care for creation.

4

Caring for Creation: The Spectrum of Belief

What does the Bible look like when we try to see it statically, as a single and simultaneous metaphor cluster? We are perhaps not too surprised to discover that there is a factor in it that will not fit a static vision.—Northrop Frye, *The Great Code*

Leaving aside the complicating variable of all the world's religions, the question before us is, "Can religion make a difference in the context of our liberal-democratic state?" And the answer is, "Yes, regardless of religious dispensation, for the Creation undergirds the very possibility of the meaning of life." Yet the enormous diversity of religious belief, let alone scientific theories of cosmic evolution and life on earth, complicates this thesis.

To begin with, the Creation does not mean the same thing to all people even *within* the biblical tradition. How are we to read the creation stories in Genesis? Clearly, the Great Code yields more than one interpretation. Conservative Christians who believe that the Bible is infallible recognize the existence of alternative if inaccurate readings. There is also variance on the Creation within the moderate to liberal range of Judeo-Christian belief. Do evolutionary accounts of life on earth supplant Genesis? Or do they call for a reinterpretation of Old Testament outlooks on creation? Further, there are religions in modern America that fall outside a biblical tradition. Goddess feminists, for example, believe that the *Magna Mater*, the Great Mother, is the divine Creatrix. And other creation stories draw on scientific models, some of them complementary to and others irreconcilable with religious narrative.

Given this variety, it is difficult to provide a definitive analysis or conduct an exhaustive investigation of either creation stories or metaphors of caring for creation. No such attempt is made here. But a survey along *a spectrum of traditions of faith* in North America, from nature and Goddess religionists on one end to Judeo-Christian conservatives on the other, is realistic. Approximately twelve hundred religions presently exist in North America; accounting for the creation story of each tradition is not feasible here. With sufficiently general categories, however, most can be positioned in relation to the biblical tradition.

This chapter does not argue that any one tradition of faith is either truer to Judeo-Christian traditions and morally superior, or more metaphysically enlightened and scientifically informed, than another—although specific traditions do claim such advantages. My primary goal is to show that creation stories across the spectrum of belief coalesce, despite their differences, around a politically efficacious—or at least potentially useful—metaphor of caring for creation. Beyond that, the survey illustrates the already discussed idea that when metaphors are properly grasped, they are appreciated as vital agents for ethical and intellectual progress (Rorty 1991b, 172). Which is to say that useful metaphors are likely more important than any other linguistic matter in a time of ecological crisis, since through them Americans might begin to reweave the modern story. Metaphors are essential to recasting our conception of the world and human behavior in it, and this all the more so when the metaphors derive from a foundational mythic structure.

In what is best understood as an enormous irony, religious discourse—which has become increasingly privatized and therefore irrelevant to public affairs—may reemerge in the public forum through environmental concern. Such renewal is essential, since the legitimating narrative of the West hangs on the idea of progress, itself derived from Judeo-Christian traditions. The modern narrative not only is incapable of sustaining but is undercutting the idea of progress: a society that undercuts the ecosystems upon which its existence depends is self-defeating. Of course, some people believe that scientific information and ecological education is enough to move society in a new direction. The paradox of environmentalism (as mentioned in the Introduction, above) makes such a conjecture enormously problematic. Science alone, as physicists like Einstein and Schrödinger and such biologists as E. O. Wilson and Stephen Gould agree, cannot direct human action. But religion can provide a rationale for behavior, and religious metaphors and symbols are especially potent influences. Just as a flag can inspire patriotic feelings and acts, so metaphors can inspire people to think and act in novel ways, in fashions that go beyond the bounds of ordinary discourse and behavior. By emphasizing the iconoclastic potential of a metaphor of caring for creation, I underscore the potential of *the second language* inherent within the biblical tradition specifically and religious discourse more generally to expand a cultural conversation about ecology beyond the language of utilitarian individualism.

In this chapter I focus on creation stories. There are obviously enormous differences across the spectrum of belief, but caring for creation is a central metaphor that *might serve* to unite all traditions of faith in setting an environmental agenda.[1] As Lakoff and Johnson suggest, "The most fundamental values in a culture will be coherent with the metaphorical structure of the most fundamental concepts in the culture" (1980, 22). If Americans can articulate a cultural core of belief to care for the creation in which all creatures live and move and have their being, a green transformation of the modern story will almost surely occur. A creation story is primordial, carrying both obligations with it and injunctions for human behavior toward all aspects of the world.

My framework for analysis (figure 1) is an artifice that allows divergent traditions of faith to be positioned relative to the Great Code. These categories are too wide, and therefore lack the connotation necessary to precise theological distinctions. Given the diversity of religion in America, they are also too narrow, a Procrustean bed into which all are thrust. And in some cases they imply denotations that are inconsistent with established usages. For example, the conventional connotation of Conservative Jews places them, in terms of these categories, as moderates. However, my interest is not theology (rigorous

Figure 1 The Spectrum of Belief

Inside the Judeo-Christian Tradition				**Outside the Judeo-Christian Tradition**
Conservative	Moderate	Liberal	Radical	Alternative

theory) nor sociology (rigorous description) nor semantics (established usage) but the political utility of religious discourse that centers on a metaphor of caring for creation.[2] Labels (or categories) also carry potentially misleading connotations. The term *radical* can be construed as "other" and therefore inferior or as "subversive" and therefore dangerous. That is not my intent. By *radical* I specifically mean traditions of faith that rise up to challenge the Judeo-Christian mainstream, often by reading the Bible in ways that reveal what they consider a deeper or more fundamental account of Judeo-Christianity. So constrained, the following distinctions are useful (though subject to exception).

Conservatives within the Judeo-Christian tradition believe that the Bible is the infallible and authoritative word of God. Conservatives advance *biblically based creation stories* exclusively: the Bible is effectively a revealed account of history. God created the earth in six days, for example, and Adam and Eve were real people.[3] Conservatives are typically theists who believe that God is radically other, that is, independent of the Creation. My study focuses on conservative Protestant creation stories and briefly examines the creation stories of orthodox Judaism and conservative Catholicism.

Moderates believe that the Bible is authoritative and divinely inspired but not inerrant. They also recognize the existence of some truth in other sacred religious texts. Moderate Judeo-Christians also believe that the Bible may be reinterpreted contextually. For example, biblically based creation stories are legitimate, but not to the exclusion of scientific sources. The Genesis story of the creation of the earth in six days is not accepted as a literal truth but as an account that can be reinterpreted consistently with creation stories based in science. Similarly, moderates might reinterpret the story of Adam and Eve in light of evolutionary theory. Moderates and Conservatives often exist within the same denomination, as with the Baptists and the Methodists. Moderates can be either theistic or panentheistic, lying between the theism of conservatives and the panentheism (God is *both* separate from or transcendent *and* a part of or immanent in the world) of liberals. Moderate laity tend toward theism, the clergy toward either panentheism or a version of theism that allows for general revelation.

Liberals remain within the biblical tradition but interpret the Bible in a variety of ways by using source criticism, discourse theory, and other techniques. The Bible is viewed as neither absolutely infallible nor as invariably literal but as a text that demands interpretation like any other text. Liberals, as categorized here, are inclined toward panentheism. There is an extraordinarily wide variety in liberal positions, reflecting the diverse possibilities of interpretation. Five representative examples are briefly discussed: Teilhard de Chardin's evolutionary Christianity, John Cobb's theocentrism, Jay McDaniel's ecological spirituality, Rosemary Ruether's postpatriarchal Christian ecotheology, and Martin Buber's I-Thou theology. Each of these positions, like those of conservatives and moderates, enables a metaphor of caring for creation.

Radical faith traditions retain some trappings of the biblical tradition while offering criticisms of some biblical beliefs, such as the idea that God made *man* in *his* image, on the grounds that they are not only erroneous but also ecologically harmful. Thomas Berry argues that postbiblical creation stories based on scientific narrative are imperative for the survival of Judeo-Christian culture. The so-called new cosmology underpins Berry's creation story. Other radicals, such as Matthew Fox, advocate a creation spirituality that reflects iconoclastic biblical interpretation. And nature religionists, such as John Muir, reflect a persistent strain of American religious belief that finds God shot through nature in continuing acts of creation. Radical traditions of faith are often panentheistic and sometimes pantheistic (God is wholly immanent in the world). Radical creation stories, insofar as I treat them here, embrace scientific narrative.

Alternative faith traditions are *outside* the Judeo-Christian faith and include creation stories that either silently ignore or vociferously reject Bible-based creation stories. Among these are Gaians, Goddess feminists, feminist Wiccans, neo-pagans, and indigenous (Native American) religions. These traditions are non-Judeo-Christian because the Bible has no authoritative stance for them. My positioning of alternative faith traditions simply *locates* rather than marginalizes them. Eastern religions, such as Hinduism, Taoism, and Buddhism might also be characterized as alternative faith traditions, but they generally lie outside the scope of this study (and also lack significant political influence in North America). Some ecophilosophies, notably deep ecology and ecofeminism, also include a religious dimension.

Few if any contemporary traditions of faith could not develop a rationale, consistent with their core doctrines, to care for creation. Some denominations, it must be noted, such as the Church of Jesus Christ of Latter Day Saints (the sixth largest denomination in the United States), have made explicit statements that ecological concerns are not part of the church's mission, although the prophet of the Mormon church could change that policy with a single edict. Other large denominations, such as the Church of God in Christ (seventh largest denomination), have not yet taken any action on ecology. But there is no reason in principle that all denominations cannot find their way to caring for creation in a way that is consistent with doctrinal claims to ultimate knowledge. In fact, a study by Marshall Massey (1991) indicates that more than 90 percent of the church-going population in the United States, including the Catholics (first, with 36.85 million voting members), the Southern Baptists (second, with 12.45 million voting members), the United Methodists (third, with 7.2 million voting members), and the Presbyterians (eighth, with 2.7 million voting members), belong to denominations that explicitly acknowledge environmental concerns (6). *The ecotheological movement is already under way.* Massey believes that "if a body of Christians begins striving in a thorough and religious way to sin no more ecologically, the community might, by ripple effect, turn a whole religion around" (5).

The articulation of ecotheological rationales moves each of these denominations in the direction of green politics and the public church (see chapter 5, below). A *continuum of ecotheologies* (figure 2) illustrates some possibilities for ecotheology that develop a metaphor of caring for creation. The continuum is neither definitive nor exhaustive and does not preferentially position any faith tradition. From the vantage point of sociolinguistics, all religious discourse—including ecotheology—is similar: its primary function is to legitimate life. More generally, a religion is alive only insofar as it makes a meaningful life possible. That is to say, religious stories endure only insofar as they are employed in human context.

A list of six common features among creation stories (table 4) supplements figures 1 and 2. The most important function of a creation story is to legitimate the present by locating it in sacred time. Creation stories generally attempt to close the gap between existence and meaning by situating human beings in a cosmic continuum that stretches from the origin of time to some unknown future. Humans find themselves, their place in the cosmic scheme, by being placed within a creation story, by being located in a context that legitimates and gives direction (meaning, purpose, significance) to existence.

Figure 2 The Ecotheological Spectrum

Inside the Judeo-Christian Tradition				Outside the Judeo-Christian Tradition
Conservative: exclusively Bible-based creation stories	Moderate: biblically supported creation stories	Liberal: scientifically informed creation stories in biblical context	Radical: Judeo-Christianity in the context of the new cosmology	Alternative: non-Judeo-Christian creation stories

Creation Stories and Scientific Narrative

Scientific discourse, though it offers what might be termed "adumbrated or restricted creation stories," does not fall along the spectrum of faith (figure 1, above), since it cannot per se provide a sacred canopy. The reason is straightforward: science does not prescribe ideals for human behavior, although its descriptions do bear meaningfully on ethical choice (if for no other reason than that offered in the *Nicomachean Ethics*: informed choice is superior to uninformed choice). Further, science does not attempt to deal with questions of ultimate meaning and significance—that is, core concerns such as death or the problem of evil. These issues are beyond the province of science since the enabling assumption of modern science is that cognizing subjects—namely, human beings as specifically human beings—are excluded. Traditional scientific narratives concern only questions of efficient cause: no question of teleology is admitted.[4] Religious narratives, in distinction from science, offer human beings a way of life. As Ian Barbour argues, religious language "serves diverse functions, many of which have no parallel in science" (1990, 88). Among these are fostering ethical attitudes and acts, evoking sentiments and feelings, and effecting "personal transformation and reorientation (salvation, fulfillment, liberation, or enlightenment). All of these aspects of religion require more total personal involvement than does scientific activity" (88).

Table 4 Characteristics of Creation Stories

- Account for a creator or creative principle.
- Offer a spatio-temporal narrative.
- Account for the origin of life and human beings.
- Define the relation of human beings to the creator or creative principle and other life-forms.
- Prescribe appropriate and proscribe inappropriate behavior.
- Explain the great mysteries or core concerns of life, such as death.

Not all scientific and religious accounts of creation, however, are intrinsically antithetical. As I have discussed in chapters 1 through 3, there is reason to think that democratic solutions to environmental problems entail both science and religion. Some scholars, such as Barbour, argue that there is an enormous potential for identifying common ground shared by science and religion.[5] Unlike natural theology, Barbour points out that a *theology of nature* (or ecotheology) starts from religious tradition and experience that is modified in light of scientific findings which bear on it, particularly in such areas as the doctrines of creation and providence. His position is best characterized as moderate or even liberal, since he believes that religious doctrine, if it is to remain cognitively viable, must be consistent with science. My position is that caring for creation does not require that everyone go this far. Though Barbour's position accurately reflects the accommodation that moderate and liberal believers have made to science, it unnecessarily marginalizes conservative creation stories. Conservatives can take ecologically responsible and scientifically informed actions consistent with their claims to ultimate knowledge.

From a sociolinguistic perspective, there are no insuperable metaphysical barriers between science and religion that preclude meaningful exchanges. Even such elementary concepts as pollution, I have argued, have both descriptive and prescriptive dimensions. Of course, there may be antipathy between, for example, a conservative creation story based on Genesis and an evolutionary creation story like Richard Dawkins's *Blind Watchmaker*. Yet, in spite of a diversity of explanation, there is a convergence across the entire spectrum of religion and science on the reality of the Creation itself. Viewed sociolinguistically, rather than metaphysically, scientific and religious discourse may reinforce each other in building solidarity on an environmental agenda.[6]

There is no reason why this cannot happen. In the past science and religion mutually supported each other: three and a half centuries ago much of the science of Newton and Descartes was consistent with the worldview of the Roman Catholic church. Protestants were even more receptive. As Ilya Prigogine notes, it is difficult to imagine the scientific revolution occurring independently of the idea that God had created an orderly and therefore comprehensible world.[7] "Classical science was born in a culture dominated by the alliance between *man*, situated midway between the divine order and the natural order, and *God*, the rational and intelligible legislator, the sovereign architect we have conceived in our own image. . . . With the support of religion and philosophy, scientists had come to believe their enterprise was self-sufficient, that it exhausted the possibilities of a rational approach to natural phenomena" (Prigogine and Stengers 1984, 51). The great irony of the first scientific revolution is that, although it could not have begun without the undergirding of religion, especially the idea that the Divine Creator had an orderly mind and created an intelligible world order by design, traditional religious belief soon came under attack.

For traditionalists the cosmos was a creation of a personal God, a God above nature who fashioned it from nothing and set it into eternal motion according to a divine plan. This personal God had also fashioned *man* in *his* image, occupying a privileged place in the divine scheme of things and having dominion over the earthly creation. And the passage of time was meaningful, for human life on earth was only one aspect of a grander sweep of cosmic events. As classical science evolved and the Enlightenment "Philosophy of Man" emerged, these traditional beliefs became tenuous. The Genesis account of creation, so important to conservatives, was dismissed by many as the mythic writing of a prescientific people. The modern creation story (naturalism) is that only impersonal forces of evolution have shaped the earth and its creatures. Further, humankind occupies no special place in the evolutionary scheme. As Darwin rather bluntly put the point, "He who is not content to look, like a savage, at the phenomena of nature as disconnected, cannot any longer believe that man is the work of a separate act of creation" ([1871] 1952, 590). From a modern scientific perspective, humankind is not created *imago dei* and history is conceptualized as sound and fury signifying nothing. What follows death is simply the corruption of flesh and the end of the person.

Theism became scientifically untenable and Judeo-Christianity was recast by liberals on the grounds of deism—sometimes termed a "halfway house" to atheism. As David Griffin notes, "The human soul was no longer understood to have daily intercourse with God or transcendental moral values, and the Bible was no longer seen as a story of God's involvement in our history" (1989, 119). In consequence, Judeo-Christianity became increasingly irrelevant in the modern social scheme. Religious narrative was pushed to the edge of modern political and economic life, speaking not so much to the common good as to the needs of the faithful.

Whatever its origins, the United States of America is now a secular state where religion seems to have little or no function in the operation of our basic political and economic institutions. William Leiss observes that "Western religion has lost its hold over the domain of practical activity, and the increasingly secularized character of social behavior renders unlikely the prospect that it may someday restore its hegemony" (1972, 197). The first language of American society is utilitarian individualism. The market is believed to be the ultimate arbiter of social choice, guiding the investment of capital, the allocation of labor and natural resources, and the application of science and technology to push society ever onward and upward. The notion that a good society might collectively pursue moral progress rather than material success for the sake of material success appears to the modern mind as a quaint anachronism. Yet, as argued in chapter 3, religious discourse has a vital role to play, at this juncture of Western history, in promoting a cultural conversation on an environmental agenda and on the limits of utilitarian individualism more generally.

A merit of a sociolinguistic perspective (that the reading and rereading of texts, that the telling and retelling of stories, defines our specifically human beingness) is that no discourse is marginalized.[8] Narratives are simply that, narratives, and not one enjoys a privileged position, not one is "totalizing." Not even scientific discourse, such as the theory of evolution, can escape its history, its social construction. More generally, there is neither a "universal we" nor a "transcendental subject" that can be appealed to in order to justify any narrative without evading the reality of language.[9] Views such as Prigogine's are fully aware of the reality that language speaks and that scientific discourse is no more and no less than a form of life, another among many language games that humans play. By thinking of ourselves as storytelling culture-dwellers the cognitive legitimacy and meaningfulness of all creation stories are recognized. Through such recognition the possibility of achieving solidarity on common goods and public values, such as sustainability, is enhanced.

Conservative Creation Stories

Protestantism

Conservative creation stories are the place to begin, since until recently most environmentalists placed the blame for ecocrisis on Judeo-Christianity. These stories relate that *man* is made in God's image, that God is transcendent, above *his* creation, and that God gave to man the power of dominion over the earthly creation.[10] An entire generation of environmentalists, following the lead of Lynn White, Jr., recite in a litany of conservationist faith the charge that the root of ecocrisis lies in the Old Testament, where God gave man dominion over the earth, and in the New Testament emphasis on salvation. Environmentalists criticize the conservative creation story because of its anthropomorphism, anthropocentrism, and dualism. They charge that the conservative belief that God gave man dominion over the earth has led to its relentless humanization and exploitation. Additionally, environmentalists argue that there is an antipathy between evolutionary stories and conservative creation stories that render the biblical creation stories cognitively meaningless. White even goes so far as to say that "we shall continue to have a worsening ecologic crisis until we reject the Christian axiom that nature has no reason for existence save to serve man" (1973, 29).

Environmentalists, then, are likely to be surprised that religious conservatives argue that *the Bible explicitly mandates that humankind is to care for creation*. Francis A. Schaeffer's *Pollution and the Death of Man: The Christian View of Ecology* is perhaps the earliest defense of Christianity against the criticism of Lynn White. While there are now more comprehensive conservative texts, such as *Earthkeeping in the Nineties: Stewardship of Creation*, edited by Peter De Vos, that integrate ecology, economics, and philosophy with conservative Christian thought on stewardship, Schaeffer's book remains unsurpassed in clarifying how the Bible story makes environmental responsibility incumbent on conservative Christians. My purpose is not served by critiquing conservative creation stories from the standpoint of another position—such as process-relational theology or post-patriarchal Christian feminism.[11] Neither is there utility in assuming conservative narratives to be definitive, thereby shifting the blame for ecocrisis to either godless humanists or capitalistic entrepreneurs. Wesley Granberg-Michaelson (1984), for example, argues that the exploitation of nature is rooted in modernism, in the secularism that led to the rise of industrial growth society and the perception of nature as standing reserve, and not Judeo-Christianity. What is useful is examining biblicist creation stories to determine how these stories might help conservative Protestants achieve solidarity on ecological issues.

Conservatives argue that Nature has a value in itself because God made it. For conservative Protestants, such as Schaeffer, man (Schaeffer's term) has a primary relationship to God, since *he* is made in God's image. Further, man has been given dominion over the earth and the creatures by God (see Gen. 1.28). But—and here is Schaeffer's key point—man has a biblically based obligation to respect and care for the Creation. "Christians, who should understand the creation principle, have a reason for respecting nature, and when they do, it results in benefits to man. Let us be clear: it is not just a pragmatic attitude; there is a basis for it. We treat it with respect because God made it" (1970, 76).

The argument that the Fall (Gen. 3) caused ecocrisis is also helpful, since that explanation helps conservatives understand ecocrisis *and* prepares the way for New Testament grounds for earthkeeping (see below). The belief that sin ruptured the right relation between man and nature is widely shared among conservatives. The consequence of the Fall is the separation of man from nature. The Fall led to sinful relations (or "deep separations," as Schaeffer calls them) between man and God, man and woman, and man and nature. Some conservatives believe that the healing of this fissure lies in the distant future, awaiting the return of Christ to earth. But Schaeffer disagrees. "Christians who believe the Bible are not simply called to say that 'one day' there will be healing, but that by God's grace substantially, upon the basis of the work of Christ, substantial healing can be a reality here and now" (1970, 67).

Interestingly, a conservative position can address the practical considerations involved in problems of global ecology, such as population growth, biodiversity, and climate heating, as well as the more traditional issues of social justice. For example, the consensus of opinion within the expert community is that the problem of greenhouse gases cannot be dealt with independent of the problem of economic inequity between industrialized and developing nations.[12] For conservative Christians the Fall has caused these social fissures. It is necessary for Christians and the church, Schaeffer argues, to work toward healing them. And it is also incumbent on Christians and the church to work toward healing the fissure between man and nature. "God's calling to the Christian now, and to the Christian community, in the area of nature . . . is that we should exhibit a substantial healing here and now, between man and nature and nature and itself, as far as Christians can bring it to pass" (1970, 69).

Following Francis Bacon's lead, Schaeffer argues that Christians are to employ both arts and sciences to heal the fissure between man and nature. "A Christian-based science and technology should consciously try to see nature substantially healed, while waiting for the future complete healing at Christ's return" (1970, 81). There are undoubtedly parallels here between Schaeffer's view and what is called "restoration ecology." But for Schaeffer there can be no true ecology (one that escapes the instrumentalist ideology of the progressive conservation movement) outside of the biblical story of creation. "When we have learned . . . the Christian view of nature—then there can be a real ecology; beauty will flow, psychological freedom will come, and the world will cease to be turned into a desert" (93). For Schaeffer the key to healing the fissure between man and nature is right or proper relationship to nature and to science. "Man has dominion over the 'lower' orders of creation, but he is not sovereign over them. Only God is the sovereign Lord, and the lower orders are to be used with this truth in mind. Man is not using his own possessions" (69).

Such a biblically based argument can be derived from a variety of biblical resources. Consider, for example, Psalm 24.1 (RSV translation): "The earth is the Lord's and the fulness thereof, the world and those who dwell therein."[13] Such a passage makes any interpretation of Genesis 1.28 as a license for the exploitation of earth to suit only human ends problematic. Psalm 24 is a call for responsible stewardship, placing humans in the role of caretakers of the earth on behalf of the owner. Just as an owner might evict irresponsible tenants, so God would be displeased by a failure to care properly for the earth, for his creation.

For conservative Christians, nature has *intrinsic value* because God made it, not simply instrumental value for human beings.[14] Although there are hierarchies in the Creation, reflected in the reality that God gave man dominion over nature, all the levels of the Creation, including inanimate nature, have value. "The Christian is called upon to exhibit . . . dominion, but exhibit it rightly: treating the thing as having value in itself, exercising dominion without being destructive" (1970, 72). Schaeffer approves of Saint Francis's theology and exhibits an almost Muir-like sensitivity toward inanimate earth. Even a rock has intrinsic value. "If you must move the rock . . . ," Schaeffer argues, "then, by all means, move it. But on a walk in the woods do not strip the moss from it for no reason and leave it to lie by the side and die. Even the moss has a right to live. It is equal with man as a creature of God" (75).

Schaeffer also addresses the question of the role of the church in a time of ecocrisis, characterizing it as a "pilot plant." He takes Christians to task for not exemplifying in practice what the Bible story teaches. Schaeffer illustrates how human greed (an unrestrained profit motive) and haste (a consequence, again, of the effort to maximize profit or of poor engineering) lead to environmental despoliation, as in the case of strip mining that leaves open, acid-oozing, life-threatening scars on the earth. Christians (just as secularists) living in a capitalist society have become enslaved to a technological imperative.[15] "We have spoken loudly against materialistic science," Schaeffer contends, "but we have done little to show that in practice we ourselves as Christians are not dominated by a technological orientation in regard either to man or [to] nature" (1970, 85). Just as males have exploited females as sex-objects, so, too, have Christians exploited nature, Schaeffer continues. But what goes around, comes around. Just as our sexual lives are in shambles, so, too, are our relations with nature. "If we treat nature as having no intrinsic value, our own value is diminished" (87).

The implications for an environmental agenda are clear. Nature must not be treated by Christians as having use-value only: the imperative of the bottom line is a false god. When modern man, Schaeffer continues, "speaks of protecting the ecological balance of nature," it is characteristically "only on the pragmatic level for man, with no basis for nature's having any real value in itself. And thus man too is reduced another notch in value and dehumanized technology takes another turn on the vise" (1970, 90). Conservative Protestants, then, have every reason to care for creation.

Since 1970, when *Pollution and the Death of Man* was published, a number of remarkable texts that extend and refine Schaeffer's arguments have been written, such as an edited volume by Calvin B. DeWitt, *The Environment and the Christian* (1991). These works are important for developing the Protestant conservative's sense of obligation to care for creation because they develop not only the Old Testament traditions but New Testament resources. DeWitt (1991) notes that some of the faithful have come "to see God's Holy Word as a kind of ecological book—a manual helping believers to live rightly on earth—a manual that would help us 'so live on earth that heaven would not be a shock to us' " (7). The New Testament, for these conservatives, makes substantial contributions to developing a Christian mission of environmental stewardship of the Creation; clearly, the Creation is not only an Old Testament but a constant biblical theme, from Genesis to Revelation.

The Environment and the Christian is exemplary because it integrates a scientifically informed discussion of specific problems, such as abuse of the land, extinction of species, climate change, and waste management, with both the Old Testament creation story and the New Testament.[16] DeWitt grounds his discussion of the human assault on the Creation with an insightful economic critique. "In the last several centuries we have chosen to redefine the long-recognized vices of avarice and greed as virtues. . . . Self-interest, we now profess, is what brings the greatest good" (1991, 22–23). What, DeWitt wonders, does the New Testament teach the Christian about caring for the earth, about overcoming the degradation that has followed the relentless economic exploitation of the creation? A great deal, as it turns out. Five contributors offer useful discussions that help integrate the Old and New Testament doctrines on creation. For example, Loren Wilkinson offers an insightful account of Jesus as Creator and Redeemer, focusing especially on a scriptural understanding of atonement rooted in New Testament cosmic Christology. Disagreeing sharply with Matthew Fox that we are all Christs and "have simply to discover it," Wilkinson insists that the New Testament "teaches that Christ graciously enables us to share both in God's immanence and transcendence. Through Christ, we represent God toward creation" (41–42). Ecological sin, as Wilkinson sees it, occurs when humans begin to think that they are God. "It is when we think that we have divine power over the rest of creation that we are likely to crush it, whether that devastation be as dramatic as the nuclear disaster at Chernobyl, or as trivial as household garbage" (42)

Catholicism

Conservative Roman Catholics, though they might find merit in arguments like Schaeffer's, introduce some complications. For one reason, conservative Catholics differ from Protestants in their relation to religious authority. Catholics do not stand as do Protestants in relation to the Bible or to biblically inspired creation stories. The pope specifically and the ecclesiastical structure of the church more generally provide authoritative pronouncements. A theologically conservative pope would, in principle, support an orthodox creation story, especially as this has been articulated by such individuals as Augustine and Thomas Aquinas. Pope John Paul II has addressed environmental issues, and other Catholic prelates, such as Father Stanley Jaki, have done so, too. Jaki's view is much like Augustine's and Aquinas's. The creation of God is good. Original sin, however, ruptured the relation between man and nature. Ecological crisis is the consequence. Only obedience to God's law can restore what sin has put asunder.[17]

For both Augustine and Aquinas the Fall created a rupture between man and the natural order. One consequence was that nature no longer yielded its bounty willingly. While man had once lived in Eden, after the Fall nature had to be subdued through work. According to Aquinas, "What was brought on him as a punishment of sin would not have existed in paradise in the state of innocence. But the cultivation of the soil was a punishment of sin" (1952, 526). Pain and the suffering of earthly existence was a further consequence of original sin. Man—the natural man or woman—was for Augustine and Aquinas intrinsically flawed: only God's grace permitted redemption. Of course, the idea that nature is fallen is foreshadowed by Paul (see Rom. 8.19–23; 1 Cor. 15.24–28; Eph. 1.9–12, 22–23). Russell argues that Paul "regarded the whole of nature as being in some way involved in the fall and redemption of man. He spoke of nature as 'groaning and travailing' (Romans 8:22)—striving blindly towards the same goal of union with Christ to which the Church is tending, until finally it is re-established in that harmony with man and God which was disrupted by the Fall" (1981, 64).

Since the pope's voice is authoritative on matters of theology, conservative Catholic laity do not require a specific rationale to take action on an environmental agenda: a papal pronouncement is sufficient. For these Catholics, the pope stands in an uninterrupted line of episcopal succession from Jesus and the twelve Apostles. And for more than a century, beginning with Leo XIII's encyclical *Rerum Novarum* (1891), popes have addressed the social implications of Catholicism—that is, the economic and social conditions that have accompanied the industrialization of Western culture.[18] In 1971, Pope Paul VI addressed environmental crisis in an apostolic letter, observing that "man is suddenly becoming aware that by an ill-considered exploitation of nature he risks destroying it and becoming in his turn the victim of this degradation" (28). The problem, according to Pope Paul, is not only material but conceptual and spiritual, since "the human framework is no longer under man's control, thus creating an environment for tomorrow which may well be intolerable" (28). He suggests that Christians must take account of these "new perceptions in order to take on responsibility, together with the rest of men, for a destiny which from now on is shared by all" (28).

More recently, and more than once, Pope John Paul II has addressed ecocrisis, most notably in his message on the World Day of Peace in January 1990 (and in a briefer statement from the Holy See before the Earth Summit in June 1992). His World Day of Peace message is a resounding statement that humankind is recklessly endangering the Creation. He calls the ecological crisis "a moral issue" and argues that a socially just, religiously inspired, ecological ethic is required. John Paul turns to the biblical witness to identify the underlying problems, primarily original sin, which led humans to violate the earth. "Made in the image and likeness of God, Adam and Eve were to have exercised their dominion over the earth (Gen 1:28) with wisdom and love. Instead, they destroyed the existing harmony *by deliberately going against the Creator's plan*, that is, by choosing to sin" (1990, 4). The pope also takes issue with critics who have charged that Christian orthodoxy encourages those who would dominate the earth. According to John Paul, human dominion over the Creation, granted to humankind by the Creator, "is not an absolute power, nor can one speak of a freedom to 'use and misuse,' or to dispose of things as one pleases. The limitation imposed from the beginning by the Creator himself and expressed symbolically by the prohibition not 'to eat of the fruit of the tree' (cf. Gen 2:16–17) shows clearly enough that, when it comes to the natural world, we are subject not only to biological laws but also to moral ones, which cannot be violated with impunity" (1987, 65).

Despite humankind's unrelenting assault on the Creation, John Paul finds inspiration for the future through Christ, since he "accomplished the work of reconciling humanity to the Father, who 'was pleased . . . [ellipsis in original] through (Christ) to reconcile himself to all things, whether on earth or in heaven, making peace by the blood of his cross' (Col 1:19–20). Creation was thus made new (cf. Rev 21:5)" (1990, 5). In short, according to John Paul, the biblical witness helps Christians realize that "the destruction of the environment is only one troubling aspect" of a profound moral crisis (6). Near the end of his World Day of Peace message, he summarizes his theological position, noting that "the commitment of believers to a healthy environment for everyone stems directly from their belief in God the Creator, from their recognition of the effects of original and personal sin, and from the certainty of having been redeemed by Christ. Respect for life and for the dignity of the human person extends also to the rest of creation, which is called to join man in praising God (cf. Ps 148:96)" (13).

John Paul emphasizes that no "society can afford to neglect either respect for life or the fact that there is an integrity to creation" (1990, 7–8). The lack of respect for life, according to John Paul, is manifest in the ways the earth's peoples pour toxic chemicals into local and global ecosystems. The "reckless exploitation of natural resources" upsets ecological integrity (7). Greed, a failure to restrain the economic impulse, is responsible for the way humankind is treating the Creation. "Modern society," he argues, "will find no solution to the ecological problem unless it takes a serious look at its lifestyle. In many parts of the world, society is given to instant gratification and consumerism while remaining indifferent to the damage which these cause" (11).[19] Ultimately, John Paul concludes, the ecological crisis is telling us something: "namely, there is an order in the universe which must be respected, and that the human person, endowed with the capability of choosing freely, has a grave responsibility to preserve this order for the well-being of future generations" (13). Christians, John Paul continues, have a special obligation, even more than those who do not share the faith, because they "realize that their responsibility within creation and their duty towards nature and the Creator are an essential part of their faith. As a result, they are conscious of a vast field of ecumenical and interreligious cooperation before them" (13).

The pope sets his concern for global ecology, it must be emphasized, in a realistic context, recognizing that the developed nations have a special obligation to the Third World. "The poor, to whom the earth is entrusted no less than to others, must be enabled to find a way out of their poverty" (1990, 10). And wealthy nations have a moral obligation to help these nations. The Third World "cannot, for example, be asked to apply restrictive environmental standards to their emerging industries unless the industrialized States first apply them within their own boundaries" (9). The so-called developing nations have responsibilities, too. For one, they "are not morally free to repeat the errors made in the past by others, and recklessly continue to damage the environment through industrial pollutants, radical deforestation, or unlimited exploitation of nonrenewable natural resources" (9). Another responsibility of every Third World nation is to see that issues of economic equity, such as land reform, are carried through.

A second difference among conservative Protestants and Catholics is that where Protestants might look to the prophetic tradition for inspiration, Catholics might look to exemplars of their faith, such as the so-called desert fathers and mothers. Susan Bratton (1993), though she is not a Catholic, argues that all Christians can find inspiration for an ecologically sustainable way of life in the monastic traditions. From Bratton's perspective, Lynn White's exhortation that Christians should emulate Francis of Assisi is uninformed, since he did not spring into existence unannounced but is better comprehended as "the product of a thousand-year tradition beginning with Antony in the inner desert and kept alive by the great monastic libraries of Europe that harbored numerous tales of wildlife and wilderness loving saints" (1988, 52). John Paul, no doubt conversant with the traditions of the church in a way that Lynn White is not, names Francis as the heavenly patron of nature, observing that "the poor man of Assisi gives us striking witness that when we are at peace with God we are better able to devote ourselves to building up that peace with all creation which is inseparable from peace among all peoples" (1990, 14). Francis exemplifies a long tradition of humans who believe that God's spirit animates all creation—humankind is merely one life-form among many. For Catholics who continue in that tradition, God smiles on all creation and creatures. Nature is not there for human exploitation.

Judaism

For conservative (here meaning Orthodox) Jews a significant difference from Protestants and Catholics lies in the sacred literature itself.[20] A conservative Jewish ecotheology is based on the Bible, but the Hebrew Bible is what Christians call the Old Testament. A conservative Jewish creation story could emphasize a number of themes in the Hebrew Bible, drawing on the priestly or prophetic literature as well as on other sources of Jewish tradition. These sources include *halakhah*, the rules and statutes found in the scriptures and their interpretations; *aggadah*, or nonjuristic literature, including proverbs and parables; and *tefillah*, prayer or liturgy (Helfand 1986, 39).

Orthodox Jews could, for example, develop the covenant tradition that is fundamental to the Pentateuch. The first explicit reference to the covenant is in the story of Noah, where several passages make clear that *the covenant is between Yahweh and all of the creation*. God said, "This is the sign of the covenant that I make between me and you and every living creature that is with you, for all future generations" (Gen. 9.12). "I set my [rain]bow in the clouds, and it shall be a sign of the covenant between me and the earth" (Gen. 9.13). "I will remember my covenant that is between me and you and every living creature of all flesh; and the waters shall never again become a flood to destroy all flesh" (Gen. 9.15). "When the [rain]bow is in the clouds, I will look upon it and remember the everlasting covenant between God and every living creature of all flesh that is upon the earth" (Gen. 9.16). And finally: "God said to Noah, 'This is the sign of the covenant that I have established between me and all flesh that is on the earth'" (Gen. 9.17). There is no sense in these passages that Yahweh assigns humankind privileges over the rest of the Creation.

Isaiah further develops the tradition of the covenant, particularly where the prophet addresses Yahweh's work of the renewal and salvation of the creation. For example, Isaiah 11.6–9 embodies an idea of responsible stewardship.

> The wolf shall dwell with the lamb, and the leopard shall lie down with the kid, and the calf and the lion and the fatling together, and a little child shall lead them. The cow and the bear shall feed; their young shall lie down together; and the lion shall eat straw like the ox. The sucking child shall play over the hole of the asp, and the weaned child shall put

> his hand on the adder's den. They shall not hurt or destroy in all my holy mountain; for the earth shall be full of the knowledge of the Lord as the waters cover the sea.

This passage implies that God's intention is for all elements of his creation to live in harmony. No single part is favored over any other, since none "shall hurt or destroy in all my holy mountain." In a modern context, humans have no right to pollute the water in which the fish swim or to destroy the rainforests that support hundreds of thousands if not millions of species known only to God.

Finally, Psalm 104, a hymn to God the creator, offers a powerful passage for developing an ethic of caring for creation. Verses 10–18 reveal God's continuing care for the earth, its processes, and its living inhabitants.

> Thou makes springs gush forth in the valleys; they flow between the hills, they give drink to every beast of the field; the wild asses quench their thirst. By them the birds of the air have their habitation; they sing among the branches. From thy lofty abode thou waterest the mountains; the earth is satisfied with the fruit of thy work. Thou dost cause the grass to grow for the cattle, and plants for man to cultivate, that he may bring forth food from the earth, and wine to gladden the heart of man, oil to make his face shine, and bread to strengthen man's heart. The trees of the Lord are watered abundantly, the cedars of Lebanon which he planted. In them the birds build their nests; the stork has her home in the fir trees. The high mountains are for the wild goats; the rocks are a refuge for the badgers.

Verse 24 carries perhaps even more powerful implications for caring for creation. For the Psalmist stands in wonder and awe of the Creation: the earth is not a standing reserve awaiting human development and exploitation. "O Lord, how manifold are thy works! In wisdom hast thou made them all; the earth is full of thy creatures. Yonder is the sea, great and wide, which teems with things innumerable, living things both small and great." Ironically, as the Creation verges on an *anthropogenically engendered mass extinction of species* (Wilson 1992), this passage might again fatefully influence the course of human events. As Clarence Glacken points out in *Traces on the Rhodian Shore*, Psalm 104 has already had a major influence in the West, that is, on physico-theology and thus modern ecology (1967, 427). Then again, there is no mystery in this. In a time of crisis a literate culture necessarily returns to its fundamental stories.

My account of conservative creation stories only touches on the rich possibilities for caring for creation inherent in the biblicist tradition. Conservative Protestants, Catholics, and Jews have already started the work, drawing from portions of the Bible relevant to their own faith tradition in fleshing out an ecotheology, that is, in fully articulating their responsibilities to care for creation. The Bible is replete with stories that bear directly on issues of the moment, such as biodiversity. The Noah story, for example, readily lends itself to an ecotheological rationale that humankind is mandated by God to protect all species on earth, if for no other reasons than God's love of his creation and the covenant itself. Biblicists should be offended by the so-called God-squad (the cabinet-level Endangered Species Committee) created during the Reagan and Bush administrations, which decided which endangered species were to be sacrificed to economic growth.[21] Manuel Lujan, secretary of the Interior and chair of the Endangered Species Committee during the Bush administration, proclaimed that "God gave us dominion over these creatures," apparently to do with whatever Lujan sees fit: he can sign the death warrant whenever economic expediency is served (Gup 1992). President Bush, the self-proclaimed environmental president, campaigned during 1992 to change the Endangered Species Act, arguing that protecting endangered species was hurting the economy (Kilday 1992). The point, however, is not to attack politicians: more than anything else, as DeWitt and other conservatives make clear, politicians represent and play to *the prevailing public philosophy of greed*. But every conservative tradition of faith can find its own logically consistent and emotionally evocative rationale to care for creation.

Moderate Creation Stories

Moderates include so-called mainline Protestants and most American Catholics and Jews. Constituting nearly one-half of the American electorate, the religious middle effectively contains a political majority. Moderate creation stories cover an enormous ground, touching up against conservatives on one side of the spectrum, to the point where they might be called *neoconservatives*, and liberals on the other side, to the point where they might be termed *iconoclastic moderates*. This diversity is reflected in Protestantism, which is often beset by in-house fighting among conservatives and moderates. Such disagreement leads to strained relations and sometimes to schism. American Catholicism, however, seems generally more moderate than conservative, such that some observers believe that American Catholicism and the Catholicism of the Church of Rome are different religions. Many American Catholics, for example, use birth control methods other than abstinence and terminate marriage through divorce.

The immediate question is whether moderates, like conservatives, can find reasons to care for creation. Conservatives, as we have seen, believe that the word of God effectively speaks to the faithful in a time of ecocrisis. We are to care for God's creation, for he made it and charged us with its stewardship. But, unlike conservatives, moderates believe neither that the Bible is accurate in every historical detail nor that it permits only one true interpretation. By definition (figure 2, above) moderates go beyond Bible-based creation stories to old-new stories. The "old," in this case, is the biblical witness. The "new" is the effort either to incorporate or to reconcile scientific discourse with the Bible story. Some moderates look to scientific narrative as a text that may be used to interpret and make plausible the truth of biblical witness. Others see science as providing information that complements but does not replace the essential truth of religious discourse. A key question for the religious middle is, as Schubert Ogden suggests, "whether the essential claims of the Christian witness of faith can be expressed in terms" that are plausible against the background of scientific knowledge (1977, 135).

Some critics outside the Judeo-Christian fold deny any possible reconciliation between science and religion. William Leiss argues that the fundamentally Baconian cast of the modern age precludes the reintegration of humankind and nature consistent with Judeo-Christianity. Modernism, he writes, "conceived as the possession of power *over* nature by the human species as a whole, is an idea which makes sense only in relation to the absolute separation of spirit (God) and nature in Judaeo-Christian theology—and thus is an idea which cannot be secularized without losing its internal harmony" (1972, 188). Others, such as Ian Barbour and Norman Gottwald, are more hopeful. Gottwald finds no reason that the Old Testament cannot offer moderates inspiration, so long as a scientific appreciation of the "cultural-material evolution of humankind" is brought to bear on reinterpretation. Traditions of faith that fail to do so, he believes, "are doomed to irrationality and irrelevance, whatever diversionary consolation they offer at the moment" (1979, 709). Barbour (1990), too, believes that Judeo-Christianity can reconcile its historically established traditions with contemporary science, thereby gaining intellectual credibility and a new context for theology.

If the three-thousand-year history of Judeo-Christianity testifies to anything, it is to the resiliency of the Great Code, for time and again it has renewed itself in the face of changing circumstances. Surely global ecocrisis, engendered by the industrial growth paradigm, portends epochal change. Just as surely, there is no a priori reason to think that Judeo-Christianity will not again prove adequate to the task. *That Judeo-Christians might fail to assume their obligations is an empirical possibility*. But there are indications that Judeo-Christianity can rise to the occasion and offer scientifically plausible and emotionally powerful rationales to care for creation. Given the diversity among moderates, my discussion is necessarily restricted to a small array of possibilities.

Protestantism

Moderates include such Protestant ecotheological thinkers as David Griffin, James Nash, and Susan Bratton, who aim to preserve the essential outline of the biblical witness: God as other; the *imago dei*; the *imitatio dei*. Each of these individuals articulates positions that are consistent with traditional Christian beliefs while being ecologically sensitive and scientifically informed. But there are clear differences among them. Vis-à-vis evolution, Bratton's creation story is on the conservative side, while Griffin's makes a rapprochement with evolutionary narrative and thus moves toward the more liberal side of the moderate category. Nash is effectively in the middle, arguing for a biblical foundationalism that goes beyond the doctrine of creation to include divine image, sin, cosmic redemption, and the church.

Bratton argues that it is time for a "new" Christian ecology that leaves the Lynn White debate behind and develops specific answers to such problems as overpopulation. However, she contends, "there is very little evidence [that] traditional Christian cosmology is inadequate or needs major revision in order to provide a foundation for a viable Christian environmental ethic" (1992a, 206). Although Bratton does not confront issues posed, for example, by the apparent opposition between evolutionary theory and the book of Genesis (how can evolution be reconciled with, for example, *creatio ex nihilo?*), she reads Genesis as legitimating the *inherent worth of nonhuman others*. "The Bible clearly states that the earth is the Lord's and that God continues to interact with creation (and to take a joy in it) whether or not humans are present" (206). Divine love (*agape*) flows quietly through the Creation, she argues, while the creatures continue to praise God. Further, on her reading of Genesis 2, God set the original pair not only to tilling but to protecting and preserving the earth. "Adam and Eve were supposed not only to till Eden but also to watch or preserve it. This implies that the tilling did not require displacement of the flora God had planted. Nor does it seem to have required displacement of animals. This suggests that we need to make room for the remainder of creation, domestic or wild" (1993, 297). Bratton, then, finds no reason to abandon the traditional creation story.

But beyond the Bible itself there is a long-standing history of exemplary individuals and communities who have revered the earth—both tame and wild. In a number of papers and books Bratton discusses these exemplars, including the desert fathers and mothers, the early Celtic monastics, the Franciscans, the Benedictines, the plain farmers from European Anabaptist traditions, and contemporary individuals such as Howard Zahniser (who was instrumental in winning passage of the Wilderness Act in 1964). "These individuals and communities," she contends, "are associated with a diversity of . . . [Christian] traditions, and they have appeared throughout the entire span of church history. . . . One would hardly distinguish them as 'cosmological experimenters' and several of them are notable for their relative disinterest in scholarly theology" (1992a, 207–8). Further, she argues, conventional environmentalist criticisms, like the charge that God's transcendence leads to the abuse of nature, do not hold. "The key in actual Christian practice appears to be not whether one considers God transcendent, but whether one expects God's day to day activity to be evident in creation. Those who expect the divine to act repeatedly through the natural world, and expect to see daily evidence of God's providence in the simple, ordinary and natural, have a well-developed respect for the non-human" (209).

According to Bratton, the historical exemplars of a Christian environmental ethic were motivated by the idea of holiness, of personal and communal righteousness in all aspects of their lives. Vis-à-vis the environment "the primary common denominator is the concept that contact with creation—either through agricultural labor or [through] wilderness spiritual practice—is spiritually beneficial, and that work in, with or for creation forwards holiness and righteousness" (1992a, 208). Thus, Bratton concludes, the primary issue facing ecologically concerned Christians today is not so much developing a new cosmology but dealing with specific issues. "Many Christian books and articles now available provide a Biblical justification for environmental care; very few, however, provide detailed discussion of specific ethical issues, such as the production of CO_2 and acid rain by the industrial nations, growing human numbers, the destruction of tropical forests, and a worldwide decline in biodiversity" (211). By testing themselves against real-life issues Christians can move beyond theology to action. "In the end, it isn't our theoretical statements, but our dedication to righteousness that will count" (213).

Bratton's *Six Billion and More: Human Population Regulation and Christian Ethics* exemplifies her position, integrating scientific accounts of population dynamics with a Christian perspective. She highlights the tension between a scientific and a Christian perspective by arguing that a strictly scientific—that is, sociobiological—point of view offers no reason for humans to practice population control: the selfish gene, seeking only to perpetuate itself, blindly leads toward an ever greater human population until the global ecosystem breaks down and civilization collapses. Bratton also points out that biologists like Garrett Hardin and Paul Ehrlich reject Christianity as a guide for population regulation.[22] Hardin contends that a Christian population ethic will actually make the damage done by overpopulation worse, keeping large numbers of reproductively active human beings alive until regional ecosystems are driven into irrevocable collapse.

Bratton argues that a Christian approach to population diverges from science-based models in a number of ways, primarily in the belief that "the individual has an eternal essence (the soul) and a relationship to God that transcends physical reality" (1992b, 111). In this sense, every human being has an absolute worth and place in the cosmic scheme of things. Further, Bratton contends that Christianity is "un-Darwinian in giving Christians obligations to those to whom they are not immediately related . . . and to the reproductively unfit. . . . Western culture, in holding both Christian principles and modern science . . . in high esteem, is automatically inviting conflict over personal and social reproductive values" (111). Among these values are considerations of distributive justice, which Bratton finds totally lacking in strictly biological models and proposals. She argues that Ehrlich ignores the issue and that Hardin's lifeboat ethics favor the interests of the rich, developed world over those of Third World poor. "The Bible, in contrast, presents justice as resting in the hands of a righteous God, and the poor as God's special concern" (118).

Bratton emphasizes that Christian approaches to the question of overpopulation must be informed by science; otherwise they are bound to fail. Some ecosystems, for example, are fragile, perhaps already stressed to the limit of their carrying capacity, and some might be habitat for endangered species of plants and animals; other areas of the world are blessed with fertile soils and ample rain, capable of providing food for populations located in less fertile locales. Further, Christians must realize that no one response to any issue will be adequate in all places at all times. In her opinion "the most serious mistake made by Christians attempting to resolve issues in population ethics is to cling to the preconception that Christianity must be pro-natalist because it is pro-

Table 5 Seven Tenets of a Christian Reproduction and Population Ethic

- Every child, regardless of parentage, is welcome in the Kingdom.
- The spiritual worth of the individual is determined not by reproductive fitness or marital status but by faith; everyone is part of the family of Christ.
- We should love our neighbors, including women and children, as ourselves.
- We should care for the widow, the orphan, and the sojourner and obtain justice for the poor.
- We should raise godly children and carry the Gospel to others.
- All the Creation and creatures are the Lord's and have been blessed by Him.
- We have a responsibility to care for both our human brothers and sisters *and* for the earth.

children" (1992b, 154). Christians need to recognize, given the enormous diversity of culture, that there is no one right answer to population issues.

Nevertheless, she continues, four nearly universal core concerns cut across all cultures, Christian and non-Christian alike. These are, briefly, the health of mother and child; access to the basic necessities of life; access to a reasonable quality of life; and reproductive freedom. Individual Christians are obligated, Bratton contends, to uphold these four principles in their own lives. If a family cannot provide adequate food for a child, then reproduction should be delayed. Reproductive decision-making, she argues, should be controlled not by government but by individuals. Individual action alone, however, is not enough. "Societal response should begin as soon as there is a deviation from these four standards for human welfare, or when there is a significant risk of such a deviation" (1992b, 156). Bratton cites seven tenets of faith that she believes should guide Christians on questions of reproduction and population regulation (table 5).

Bratton's approach to population questions, then, is guided by the four standards for human welfare (which she considers noncontroversial), the seven tenets of faith, and relevant scientific information. Christians, she argues, have a biblically based obligation to care for the land: environmental stewardship is not an optional but a categorical responsibility. Since unlimited population increases and materialistic ways of life threaten the land and its creatures,

Christians need to ask themselves some difficult questions—questions that involve the deepest tenets of their faith. "Are we producing so many progeny that in rushing to provide for them, we are destroying the land? Will our children find that we have wasted their inheritance while we were trying to make a better world? How are we treating God's creatures? Are we stealing their blessing?" (1992b, 167). Bratton is candid: such questions are not easily answered. The earth's suffering grows worse and human demand shows no sign of relenting. The land, and consequently habitat for God's creatures, is being overwhelmed by an ever growing tide of hungry human beings from the Third World and an ever growing appetite for consumer goods in the First World. But—and here is the essential point—*Bratton sees no chance that the needs of the earth and humankind can be reconciled outside a Christian approach that gives people both reason and motive to respond.* As an ecologist, she appreciates the role that science plays in grappling with population growth. As a Christian she affirms that science alone is not enough.

James Nash is a Methodist who serves his church as the executive director of the Churches' Center for Theology and Public Policy in Washington, D.C. Other moderates who represent other denominations might serve to illustrate the potential for caring for creation inherent in mainline Protestantism. The Presbyterian minister Richard Cartwright Austin (1987), for example, has written a four-volume work on environmental theology. But Nash's work is exemplary for several reasons, not the least of which is his commitment to spanning denominations in *Loving Nature: Ecological Integrity and Christian Responsibility*. His arguments are readily accessible to lay-people, particularly since the text is not theologically adventurous. Nash develops what he calls firm foundations for an ecological Christianity out of traditional doctrines of creation, covenant, divine image, incarnation, and spiritual presence. And finally, and beyond almost any other ecotheological thinker, Nash develops the political implications of ecological Christianity.

Nash's treatment of an ecological Christianity, though thorough, does not aspire to become a *systematic ecological theology* (nor does he try to develop a political theology). He argues that ecologically responsible Christians do not have to abandon their faith in favor of either nontraditional or radical themes. "What is required, however, are reinterpretations, extensions, and revisions, as well as cast-offs of cultural corruptions, in ways that preserve the historic identity of the relevant Christian doctrines and yet integrate ethical insights and ecological data" (1991, 94). Nash emphasizes that faith in God as the maker, the creator of Heaven and Earth is required as a foundation for a Christian environmental ethic. He quotes Psalm 24.1 to that effect: "The earth is the Lord's and all that is in it, the world and all those who live in it."

Of particular interest is Nash's reading of the doctrine of creation. He argues that the Bible makes it clear that the entire creation, flora and fauna, is endowed with an intrinsic moral significance. There is no nature-grace dichotomy in the Bible, he argues, although this fact "has not deterred Christian churches from restricting in practice the scope of grace to matters of personal salvation, and the means of grace to ecclesiastical functions" (1991, 95). But that interpretation is, in Nash's opinion, too narrow, too restrictive, too anthropocentric, since the entirety of creation "is an expression of grace, that is, God's free and faithful loving kindness that characterizes God's nature and acts. God *is* love. The creative process, therefore, is an act of love, and its creatures are products of love and recipients of ongoing love (cf. Ps. 136:1–9)" (95). Further, Nash contends (100ff.), the Creation itself has a redemptive purpose: this places an ultimate obligation on Christians to become stewards of the earth.

Thus environmental activism follows necessarily, according to Nash, for Christians who grasp their responsibility to care for creation (1991, 193). Politics, for the ecologically committed Christian, is not about power. Instead it is "an essential means for realizing the desirable"—that is, the ethically good, the goals that affirm the Creation. In Nash's view, politics per se is not intrinsically evil, but a politics that serves only narrow economic ends or private interests is a perversion. Granted, there is a separation of church and state in the United States, but this precludes only the establishment of any religion as a state religion. God and Christ cannot be locked out of politics. "To be in communion with God the Politician, this 'lover of justice' and 'prince of Peace,' is to struggle to deliver the community of earth from all manner of evil—private *and* public, personal *and* social, cultural *and* ecological, spiritual *and* material" (193).

Nash qualifies his argument at some length, pointing out that his call for political activism neither entails the creation of a theocratic state (the Christianization of society) nor precludes the possibility of disagreement among Christians on specific issues. What is crucial is that all Christians accept their political obligations. "An apolitical posture on contemporary ecological concerns," he argues, "is righteous irrelevance" (1991, 194). Unlike Bratton, Nash does not focus on specific strategies, offering instead what he calls middle axioms or general guidelines. Of course, as Christians move from these generalities toward specific policies, issues become more and more difficult; population issues alone (as Bratton's discussion makes clear) indicate the difficulty of devising political solutions. Further, Nash writes, "Christian political activity

Table 6 Six Political Goals for Ecologically Responsible Christians

An ecologically sound and morally responsible public policy:

- must continue to resolve the economics-ecology dilemma;
- will include public regulations sufficient to match social and ecological needs;
- will protect the interests of future generations;
- will protect nonhuman species, ensuring the conditions necessary for their perpetuation and ongoing evolution;
- will promote international cooperation as an essential means to confront the global ecological crisis; and
- will pursue ecological integrity in intimate alliance with the struggles for social peace and justice.

Source: Adapted from J. Nash 1991, 199–221

must be tempered by the realization that no political posture, party, or platform can adequately represent a Christian ideal" (196). Nash sees six political goals (table 6) as obligatory for all Christians "in loyalty to Christ who seeks to minister through all humanity and in all contexts to all the needs and rights of all creatures" (194).

The theologian David Griffin provides an interesting contrast among Christian moderates. He articulates an old-new creation story, consistent with a traditional cosmology and with scientific narrative, yet distinct from the creation stories of either liberal or radical Christians. Unlike Bratton, who sees the traditional cosmology as an adequate underpinning for dealing with environmental problems, Griffin believes that science and religion must be reconciled, especially since secularism—the industrial growth paradigm, dedicated to the exploitation of the earth in order to fuel a materialistic way of life—feeds off uncertainty. He acknowledges that liberal theology attempts to reconcile faith and reason but believes that such attempts fail because they abandon the essence of the Judeo-Christian tradition. When God is denied identity as an agent, as a traditional theistic, personal God, he becomes something vacuous, like being itself. As an alternative, he suggests, Judeo-Christians might reconcile God as creator and evolutionary process by arguing that "the direction of the evolutionary process is rooted in a cosmic purposive agent" (1989, 70).

Many problems of definition, even controversy, swirl around the idea of

God, but Griffin suggests that Judeo-Christians have associated at least seven attributes with the idea. "God is (1) the supreme power, (2) the personal, purposive creator of our world who is (3) perfectly good, (4) the source of moral norms, (5) the ultimate guarantee for the meaningfulness of human life, (6) the trustworthy ground for hope in the ultimate victory of good over evil, and (7) alone worthy of worship" (1989, 77). Distinguishing what he calls *naturalistic theism* from supernaturalistic theism, Griffin argues that evolution and God can be reconciled if we recognize that "the normal causal relationship [between] God and finite beings is a natural, necessary relationship" (77). Like the conservative's God, this God has supreme power; but unlike the conservative's God, the power is limited. "All events occur through the co-creativity of God and the creatures. Divine causality is therefore always persuasion; it can never be manipulation or unilateral fiat" (77).

Naturalistic theism, Griffin believes, enables an old-new creation story that is true to the Christian witness. From the perspective of naturalistic theism, evolution is not an embarrassment to the believing Christian but an enhanced or richer, more detailed account of God's creation, the continuing relation of the divine to the ongoing processes of reality, and of our human place in it. Arguing for what he calls *prevenient grace*, Griffin explains that once Christians realize that God is with the Creation, then it follows that God calls them to evolutionary possibilities, calls Christians "not simply to accept actuality but to respond creatively to it. God especially calls us to respond to our actual environment in terms of those possibilities through which truth, beauty, goodness, adventure, and peace will be embodied and promoted" (1989, 123). When Christians recognize that God has continuing agency in the Creation, environmental issues take on a new urgency and importance. For example, a Christian would find the destruction of species and the overall devastation of tropical rainforests an abomination, for these not only diminish some of the actuality of God but begin to foreclose future possibilities. And Bible stories take on expanded meaning in the context of the environment. Just as Jesus, for example, abominated the moneylenders in the Temple, so those who recklessly exploit the rainforest, irrevocably destroying habitat and driving countless species into extinction, are affronting God.

Catholicism

Moderate American Catholics are like conservative Roman Catholics in some ways. They are unlikely to challenge, for example, basic church doctrine on the Virgin, on the role of the pope, or on orthodox teachings in the tradition of

Augustine and Aquinas. But North American Catholics do weigh the church's pronouncements in light of their own cultural circumstances; for example, they practice birth control using methods other than abstinence. As with Protestants, moderate Catholics are also likely to accord science a large role in their thinking on creation. A meeting of American bishops in Washington, D.C., in 1991 led to the approval of a detailed statement on the environmental crisis in light of Catholic social teachings. The bishops' position, entirely consistent with the statements of Pope John Paul II, is an authoritative statement that complements the church's doctrine on creation with insights from secular environmentalism and ecological science. "We find much to affirm in and learn from," the statement reads, "the environmental movement: its devotion to nature, its recognition of limits and connections, its urgent appeal for sustainable and ecologically sound policies" (United States Bishops' Statement 1991, 431). Of course, there are differences between secular environmentalism and Roman Catholicism, not so much in goals, such as sustainability, but in the rationale for change. The bishops emphasize Paul's proclamation to the Athenians that the Creator "made from one the whole human race to dwell on the entire surface of the earth, and he fixed the ordered seasons and the boundaries of their regions, so that people might seek God, even perhaps grope for him and find him, though he is indeed not far from any of us" (Acts 17.26–27). There is, the bishops stress, a goodness in the world, and "the Western mystical tradition has taught Christians how to find God dwelling in created things and laboring and loving through them" (431). The statement also underscores the importance of Christian love in environmentalism, noting that it "forbids choosing between people and the planet. It urges us to work for an equitable and sustainable future in which all peoples can share in the bounty of the earth and in which the earth itself is protected from predatory use" (431).

Moderate Catholics do not go as far with science as do liberals, such as Teilhard de Chardin, who attempts a fundamental reorientation of the creation story. But the bishops' statement on "Renewing the Earth" confirms that they are open to science. "We ask scientists, environmentalists, economists and other experts to help us understand the challenges we face and the steps we need to take. Faith is not a substitute for facts; the more we know about the problems we face, the better we can respond" (United States Bishops' Statement 1991, 432). While secular environmentalists like to charge that religion encourages anthropocentrism, thus encouraging humankind to assume that nature is no more than a stockpile of resources for human consumption, the bishops' statement affirms that natural ecology and human ecology are one.

"The web of life is one. Our mistreatment of the natural world diminishes our own dignity and sacredness, not only because we are destroying resources that future generations of humans need, but because we are engaging in actions that contradict what it means to be human" (426). So stated, an ecological perspective checks human egotism, thus complementing the pope's directives to respect the integrity of the Creation.

There is a seeming antithesis between the scientific idea that humankind is a part of a global ecological process and the orthodox teachings that nature is fallen and humankind's true place is in heaven. Since the first and second scientific revolutions, religionists (with the exception of conservatives and some moderates) have either ceded the creation story to science or modified their creation stories in light of science. Accordingly, the story of salvation, the last bastion of religious authority, has become perhaps too important for Judeo-Christians. Yet the Bible supports the notion that all the flora and fauna as well as humankind are part of the Creation and that the Creation is good. Arguably, then, there is no reason why the theme of creation must be subordinate to the theme of salvation. If Catholics reconcile scientific and religious discourse, the claim that the Christian narrative is the one true story of the universe is not undercut but enhanced. The moderate, in other words, can use scientific discourse to buttress the belief that human beings are part of God's ongoing creation—one expression of divine will. As John Haught puts the point (in a somewhat different context), "Evolutionary thinking [science] has made it possible for us to recover in a dynamic way the ancient intuition of the cosmos as primary creative subject" (1990, 171).

The intellectual path is one way to emphasize the obligation of caring for creation. But the combination of prayer, sacrament, and ritual has been the road upon which many of the faithful have traveled. Many Catholics find a better sense of their relation to the Creation through the sacraments rather than through theology. As the bishops' statement notes, "The sacramental life of the church depends on created goods: water, oil, bread and wine" (United States Bishops' Statement 1991, 431). The sacraments open the door to an awakened sense of the mystery and wonder of the Creation and encourage a respect for the earth and its creatures, since all are equally a part of God's work. The sacramentalist perhaps escapes the nominalism of conservative Catholicism, which insists on the complete separation between God and the Creation. A sacramental theology involves at its most basic level "the perception that items of material reality—water, bread, and wine—can be given a new meaning and status by being brought within the saving action of God in Christ. . . .

The true potential of bread, for instance, is revealed by its transformation into a means of communication with God" (Habgood 1990, 47). Perhaps more than anything else the eucharist confirms for the believer that the world is more than the mundane descriptions assigned by common sense, science, economics, and politics. As the bishops observe, "The whole universe is God's dwelling. Earth, a very small uniquely blessed corner of that universe, is humanity's home; and humans are never so much at home as when God dwells with them" (United States Bishops' Statement 1991, 428).

Perceived as belonging to God, the earth stands transformed. An animal or a plant, a tree or an entire ecosystem, is no longer a mere object awaiting human use or appropriation (although it may be used or appropriated) but is perceived as a being that reveals the generosity and care of the God-who-is-there. The flora and fauna, mountains and wetlands, wind and water are, as the ancient Psalmist acknowledges (Ps. 104), manifestations of the wonder and mystery of God. For the sacramentalist, "There is a respect due to them, an awareness of human limitations, a fine balance to be struck between penitence for what we have done to God's world in the past and hopeful creativeness for the future" (Habgood 1990, 50).

Finally, moderate Catholics might find inspiration in the monastic traditions, such as that of the Celts, of the Catholic church. As the Anglican David Adam points out, from its beginnings the Celtic church rested on the belief that through immediate knowledge of the Creation one also knew God the creator. The modern Christian has lost that sense of the sacredness of nature. "Much of our insensitivity to the world, our misuse of its resources, our destruction of great areas, is because we have lost awareness of the mysteries and the strange bonds that link all things. Those who only use the world for what they can get out of it have lost touch with its Creator" (1987, 88). Adams finds in Saint Patrick an avenue toward a reawakening of the sense of nature as God's work. The Celtic tradition overlaps with the sacramental vision, since all things, including the prosaic activities of human life, such as sailing a ship or nurturing a child, are seen as within the encompassing being of God. According to Adam, "Only our own dullness and blindness . . . stops us seeing that all of life is consecrated. We need to learn to look deeper and discover again that 'in Him we live and move and have our being' " (89).

Judaism

Moderate (here meaning Conservative) Jews likely have more in common with their conservative (orthodox) and liberal (reformed) brethren than with

Christians per se. As noted, Orthodox Jews find ample resources in the Torah for defining their relations with nature. But moderates go beyond the Old Testament. For example, a two-part series in the *Melton Journal*, "Towards a Jewish Ecological Paradigm: Essays and Explorations," draws on an array of non-Jewish materials, such as Paleolithic cave paintings and nature writing, to flesh out a creation story. The contributors to the series do not repudiate but rather elaborate on the traditional Jewish reverence for the land.

Central to Judaism is monotheism: there is but one God. Judaism also, on Arthur Green's (1992) view, is monistic: "For Jews God is absolute Oneness" (4). All that is, was, or will be is God. There is no thing (nothing) outside of God. But human beings perceive God in two ways. "One aspect of the divine Self," as Green explains, "is faceless, unknown and unchanging." This is the transcendent face of God, the God who is eternal, immutable, beyond the world of nature. "The other aspect of the Divine fills all the world; everything in the world is a face of God." This is the God immanent in nature, bound up with the mysterious cycles of life and death, birth and maturation. "This God is parent, mother and father. It is the God of the Kabbalist" (4). The traditional language of creation, particularly that associated with rabbinic Judaism, Green continues, inadvertently favors the transcendent face of God and thus makes God seem apart from rather than a part of the world. "Rabbinic Judaism lost [the] . . . sense of the divine presence in the natural order, or nature itself attesting to God" (5). However, Green argues, other Jews, most of them involved in the Zionist movement, such as Aaron David Gordon and Rav Kook, provide a language for creation through which the other face of God appears. They emphasized reestablishing the traditional Jewish bond with the Creation, with earth itself. These Jews "were creating a new generation and a new life for the Jewish people" (5).

Yet the Jews of the Diaspora, such as those living in North America, remained aloof. "Were we going to teach the religion of *Yidiat Ha-aretz*, knowing and loving the different kinds of stones and wild flowers in the Negev, here in America? That made no sense in our context." Today, Green argues, that position is neither scientifically plausible nor ethically defensible, since the world is locked in ecocrisis. All Jews must ask themselves what it means to be a religious person. The answer, according to Green, is that "we must *first* say . . . that we live in a created world, or we live in a God-infused world. The way we will see that is by going out to nature and opening our religious eyes to it" (1992, 5). This is not iconoclasm, he argues, but rather a revitalization of the Jewish tradition, the ancient tale that "sees each creature as a bearer of the divine presence."

> Only as we learn this will the world survive the onslaught of humanity. Only as we learn this, I dare say, will our humanity survive the onslaught of life in the world. And only as we learn this, will the presence of God shine through the world to enlighten our humanity. Only this level of knowing God will bring us to harmony, allowing us to renew and to continue our vision of oneness. This is the vision which ends, as do our prayers, with the unitive cry: *bayom hahu yehiyeh Hashem echad u'shmo echad*—on that day will God be one and the divine name one. [5]

Liberal Creation Stories

Liberals are committed to the idea that *special revelation* and *general revelation* are complementary, that each enriches the significance of the other. For a liberal (Catholic, Jew, or Protestant), science is part of general revelation—that is, human interpretations of the divine. Liberal creation stories attempt to incorporate or place scientific narrative in the context of the broader framework of religious discourse. Unlike a conservative or even a moderate, then, a liberal rereads the Bible in the context of evolutionary narratives. Genesis, for example, is not natural history but a special revelation that takes on deeper meaning through its relation to the ongoing process of general revelation.

Catholicism

Pierre Teilhard de Chardin exemplifies liberal Roman Catholicism, since he abandons a traditional creation story that hinges on being and substance (a Thomistic account) in favor of an evolutionary narrative that pivots on becoming and process. Teilhard was psychologically positioned, as both priest and scientist, at the fault line of modern culture: the fissure between science and religion. His works, among them *The Phenomenon of Man*, *The Divine Milieu*, and *Hymn of the Universe* collectively argue that Judeo-Christians need not reject their faith in order to reconcile themselves intellectually with evolution. His creation story is post-Darwinian, an evolutionary account that goes beyond the Bible and established church doctrine.

Cosmic evolution begins for Teilhard with the "primal dust of consciousness," a useful metaphor that eschews dualism in favor of monism. From the outside, that is, from the perspective of the natural scientist, matter is brute, unconscious, existing only in external relation to other material objects. From the inside, however, from the perspective of the differing elements of creation, matter has a "conscious inner face." "Refracted rearwards along the course of evolution, consciousness displays itself qualitatively as a spectrum of shifting shades whose lower terms are lost in the night" (1959, 60). The evolutionary process is governed by "the law of complexification," as manifest in the increasing order of the cosmos, created over eons of space-time. The law of complexification works—from undifferentiated clouds of hydrogen, through stellar and galactic evolution, to the local vicinity of space-time and the emergence of life—through convergent integration: that is, higher order emerges from what is relatively a more primitive level of organization. From hydrogen come stars and galaxies; from stars come later generations of stars and the elements. From these elements, subject to the inexorable reality of the law of complexification (just as matter is subject to the law of gravity), ultimately come life and a level of being that Teilhard calls the *biosphere*.

Humankind is part of this process of cosmic evolution and, consistent with the Judeo-Christian creation story, a *special creation*. For the law of complexification creates in the human brain a material system that is simultaneously self-conscious. The proliferation of human culture across the planet—what Teilhard calls the *noosphere*—manifests this extraordinary evolutionary development. Just as the biosphere is a thin living layer of organisms covering the face of the planet, so the noosphere is a thinking layer, a living dimension of specifically human meanings, created across a relatively short span of time (compared to the longueurs of cosmic evolution). "Psychogenesis," as Teilhard puts it, "has led to man. Now it effaces itself, relieved or absorbed by another and a higher function—the engendering and subsequent development of the mind, in one word *noogenesis*" (1959, 181). But evolutionary process has not reached its end in the evolution of the human species, the profusion of culture on planet Earth, and the noosphere. Teilhard argues that humankind is moving toward a single world culture and that, as a parallel development, a spiritual concentration (higher level of complexity and integration) is occurring as well: cosmic evolution is heading toward the Omega Point.

Critics argue that Teilhard's notion of the Omega Point is scientifically tenuous and philosophically vague. Yet his attempt to reconcile evolutionary science with the Judeo-Christian creation story is courageous. Omega is not God, but more a manifestation of the divine will insofar as it determines the direction and terminus of cosmic evolution. For Teilhard the march of evolutionary process toward greater complexity is incomprehensible without this divine teleology. Christianity also figures in Teilhard's vision of Omega, since *agape* (love) is an essential force moving evolution toward Point Omega. Love constitutes the divine milieu, the spirit of Christ working through natural process. As Teilhard puts it, "Love in all its subtleties is nothing more, and nothing less, than the more or less direct trace marked on the heart of the element by the psychical convergence of the universe upon itself" (1959, 265). Human personality itself is a consequence of love, part of the working out of cosmic teleology. The Christ himself, the incarnation of God in human flesh, serves not so much to atone for the sins of humankind as to motivate evolution toward Point Omega, the unification of the Creation with God. So grasped, creation and redemption are one and the same process.[23]

Teilhard does not explicitly develop the implications of his creation story for ecotheology, but it is not difficult to see the lines along which it might develop. For Teilhard, the anthropogenic destruction of life on earth testifies to a "cosmic blindness," a failure to grasp the human purpose in the grander scheme of things, the divine milieu. Although to the casual reader this may appear to privilege humans over the rest of the Creation, expositors of Teilhard's work argue that this is not so. According to James O'Brien, in Teilhard's theology human beings remain "at home in the world biologically, psychologically, and spiritually" (1988, 346). And although Teilhard lacked such ecological information as knowledge of the adverse effects of CO_2 on the atmosphere, O'Brien argues that any consistent interpretation of Teilhard recognizes an ecological ethic. "The environmental ethical limits are set by the good of the whole of which man is a part" (346).

Protestantism

Although they represent only a small sample on the Protestant side of liberal creation stories, John Cobb, Jr., Jay McDaniel, and Rosemary Ruether are prominent thinkers. Each argues, in one way or another, that life is a central—perhaps the fundamental—religious symbol for God. Judeo-Christians who destroy the earth are also killing God, surely the ultimate apostasy, since "God's life depends on there being some world to include" (Cobb 1990b, 197). Cobb argues that traditional forms of Christianity "worship God in such a way that our attention is withdrawn from our interconnectedness with the whole creation and is focused only on our own inwardness before him" (1983, 90). Liberals like Cobb do not argue that the world is God and God is the world. Nor do they argue that God cannot exist without the world (although the world cannot exist without God). The key point is, according to Cobb and Birch, that "God includes the world and the world includes God. God perfects the world and the world perfects God. There is no world apart from God, and there is no God apart from some world" (1990b, 197). On this view, whatever else God might mean to liberal Christians, God is necessarily the inclusive whole. Accordingly, all creation must be treated with respect.

Cobb's work is exemplary in liberal ecotheology, partly because, like Schaeffer among conservatives, he was among the first to respond to the criticism leveled by Lynn White. Unlike Schaeffer (now deceased), who did not develop his initial position, Cobb has written on the subject for two decades, beginning in 1972 with *Is It Too Late? A Theology of Ecology* and continuing toward the present with *For the Common Good* (1989, written with the noted economist Herman Daly) and *Sustainability: Economy, Ecology, and Justice* (1992). Cobb's work is exemplary, too, because he raises the difficult issues that ecocrisis poses for Christianity. How culpable is the Judeo-Christian story for ecocrisis? Can Christians embrace evolution? Can they encompass the biocentrism of an Aldo Leopold? or the geocentrism of a Thomas Berry? Can Christians find inspiration in the New Testament and also care about the earth? At one time, Cobb confesses, he despaired of finding answers to these questions, tending to believe that a "new Christianity," or at least a "deeply repentant Christianity," was needed (1992, 2). Now he believes that responsible Christians can answer all these questions affirmatively through a modified Christianity that turns to the Bible with new eyes. "In many cases," Cobb argues, the scriptures are "far less dualistic than their standard interpretations, sometimes even their translations." Thus, rather than seek a new Christianity, today Cobb believes that Christians, more than anything, need "a recovery of our own Jewish heritage" (4).

Cobb finds the biblical witness entirely consistent with the reality of evolutionary process and the insights of ecology. He affirms the ecological judgment that human beings are part of a natural community of life on earth and the evolutionary judgment that the community of life has evolved over tens of millions of years. This sense of community is fundamental to Cobb's ecotheism. When the human species, in its processes of economic development, endangers an ecosystem or drives a species into extinction, it has sinned. "Throughout the Jewish scriptures the land is God's, not a commodity to be owned by human beings. All the creatures are seen as praising God. In Jesus' teaching God cares for the sparrows and for the lilies of the field" (1989, 385). There is, then, no inherent contradiction between scientific and religious discourse.

True enough, Cobb admits, "modern Western Christianity has often allied itself to anthropocentrism and the neglect of nature. It has generated suspicion of organismic views of human beings and of their communities, and fear that the distinctiveness of human beings, both their specialness by virtue of having been created *imago dei* and their radical sinfulness, which distinguishes them from the rest of creation, will be obscured" (1989, 376–77). But there is nothing intrinsically wrong with Judeo-Christianity, nothing inherently at odds with a *theocentric vision*—that is, a view of the Creation as one divine, evolutionary continuum of which humankind is a part. Indeed, in part by returning to the biblical tradition, such as the prophetic tradition, Christians can begin to reform their faith. Nature does not exist merely as standing reserve for human appropriation. The truth, according to Cobb, is "powerfully affirmed in the biblical origins of the prophetic tradition that we are persons-in-community, that there is no genuinely human life when community is destroyed" (385).

The Old Testament prophetic tradition helps Christians overcome the unbridled power of economic discourse and its devastating effects on the biosphere. According to Cobb, when humankind sets economic goods against the interests of the community of life on earth, it falls into idolatry—it fails to heed the teachings of the prophets. The canonical prophets, he continues, "prophesied primarily against their own community, Israel. . . . The absolute authority of human community was exposed as idolatrous" (1989, 383). The favoring of the Christian's own narrow economic interest over a concern for the Creation is, in Cobb's opinion, akin to the worship of Mammon, a false god. He calls on Christians to take the lead in making economic reforms that lead toward sustainability. "Only when we see that our real economic needs can be met more adequately with quite different economic practices will we make the changes needed to avoid worse and worse catastrophes. . . . As Christians we are called to lead in envisioning a more livable world" (1992, 5).

The New Testament also offers a liberating message, for "the immediacy of God frees people from absolute worldly loyalties in order to bring about justice and righteousness within the world" (1989, 386). Granted, Christians place a strong emphasis on redemption. But the message of redemption, Cobb argues, does not imply that the world is merely a vale of tears from which the Christian hopes to escape. "Human beings are part of the creation, and creation is the context within which redemption occurs. Furthermore, although the story of redemption focuses on human beings, it does not exclude the rest of creation" (1992, 83). Cobb finds nothing in the New Testament that necessitates any radical, dualistic separation between human beings and nature, although some Christians have historically made that interpretation. "Jesus' insistence that a human being is worth many times as much as a sparrow has meaning only because of the affirmation that God cares about sparrows too.

Paul's teaching about redemption includes the redemption of the whole of creation. The Bible locates human beings squarely within the natural world, despite its special emphasis upon them" (92–93). In Cobb's opinion, an anthropocentric vision of Christianity that places human values over all the rest of creation is not biblically grounded. Similarly, a biblical perspective is "distorted when God is so separated from the world that the service of God can be separated from the service of fellow creatures. . . . To say that we love God when we do not love God's creatures is to lie. In serving the least of our fellows we minister to Christ" (93).

Cobb offers a theocentric vision as an alternative to the anthropocentrism of the modern age (and as an alternative to secular biocentrism). Theocentrism differs from traditional theism in that God is the inclusive whole, sum of all that was, is, or will be. Cobb claims several advantages for theocentrism. One is that theocentrism reinforces the prophetic tradition in checking idolatry. In the modern age the GNP has become a secular absolute. By renewing the human sense of the sacredness of creation, theocentrism checks the human tendency to think of nature as a mere resource to fuel the economy. Theocentrism also transcends reductionism and atomism in favor of holism and systemic thinking. The modern age encourages human beings to think of all things as related only externally. But for the theocentrist "the world is as God knows it because God's knowledge is God's undistorted inclusion of all things. God knows and values each sparrow and knows and values each human being as well" (1989, 397). Both are part of an inclusive divine whole: the Creation. According to Cobb, theocentrism also evokes emotion and directs commitment, unlike abstract, systems-level theory, such as the Gaia hypothesis. The theocentrist locates value in each member and in each relation between the members of creation. Finally, Cobb argues that the theocentrist has a commitment to the future of life on earth. "God is everlasting, and future lives are as important to God as present lives. To serve God cannot call for sacrifice of future lives for the sake of satisfying the extravagant appetites of the present" (398).

Jay McDaniel presents an interesting variation on Cobb's theocentrism, in part by positioning himself between Cobb, on the liberal side, and Thomas Berry (see below), on the radical side of the Christian spectrum. In his foreword to McDaniel's *Earth, Sky, Gods, and Mortals*, Berry notes that McDaniel, like himself, develops an *ecological spirituality*. The primary difference between them is simply that McDaniel finds a utility in Judeo-Christian narrative traditions that Berry does not.[24] Like both Berry and Cobb, McDaniel believes that Judeo-Christians need a new vision: he calls this vision an "ecological" rather than a "substantialistic" perspective. From an ecological perspective, God is not some cosmic policeman whose function is to reward the faithful and punish the sinner. Rather, "God is a nurturing self in whose ongoing life the world is included. The universe itself, with its tragedy and its joy, is God's body. This means the suffering of Jesus on the cross did not simply *represent* the suffering of God. It *was* the suffering of God, as is any and every instance of suffering" (1986, 208). McDaniel argues that an ecological approach is fully consistent with Christianity in that both God and Christ remain central. "In Jesus," according to McDaniel, "so the ecological Christian claims, we find glimpses of that unbounded love which is characteristic of God's apprehension of the world, and to which, following Jesus, we ourselves strive to be open in our own finite and limited ways" (1990, 31). But the agenda for ecological Christians is perhaps larger than for other Christians. "Stated simply, an ecological spirituality will be open to, celebrative of, and transformed by, plurality—the sheer diversity of different forms of life, human and nonhuman. It is a Way that excludes no ways" (31).

The differences between Cobb and McDaniel lie primarily in how heavily McDaniel incorporates such themes as liberation theology, Buddhism, postpatriarchal Christian feminism, and discourse theory into his work. Cobb, of course, does not ignore these.[25] But McDaniel appears to integrate them, particularly in his *Of God and Pelicans: A Theology of Reverence for Life*, more completely into his version of ecological spirituality. Rosemary Ruether, who is discussed below, in particular influences McDaniel's vision of a postpatriarchal Christianity, although he also draws on other Christian feminists, such as Catherine Keller, and ecofeminists, such as Karen Warren. McDaniel is especially attentive to the nuances of language, such as the word *God*, which many feminists (such as Ruether) react to as laden with problematic, even sexist connotations. He suggests *Heart* as an alternative for naming the divine mystery. Heart, he explains, is a word rich enough to complement "more personal metaphors such as 'Mother,' and 'Father,' 'Lover' and 'Friend'" (1989, 142). *Heart* is an alternative for women and men who find traditional religious language oppressive. Perhaps, McDaniel suggests, the divine spirit is best conceived not as "a father who art in heaven" but relationally. Heart, he argues, "is the ultimate expression of relational power. It is potentially the most influential power in the universe though the efficacy of its influence depends on worldly response, and it is the most vulnerable power in the universe. In the latter respect Heart feels the feelings of living beings, suffering their sufferings, enjoying their joys, sharing in their destinies in ways much deeper than we can imagine" (144).

Liberal Protestant creation stories also include Rosemary Ruether's. She is a postpatriarchal Christian feminist who offers alternatives that may be reflected but are not fully developed by others. Ruether attempts to integrate feminist theory and ecological insight in a way that amends the biblical tradition. Just as Cobb sets the biblical tradition within an ecological framework, so Ruether places Judeo-Christian narrative within the context of feminist narrative—she invites Judeo-Christians into the hermeneutic circle. Every great religious tradition, she argues, begins in revelatory experience, and Judeo-Christianity is no exception. Formative groups or communities (the tribes of Yahweh, early Christians) mediate "what is unique in the revelatory experience through past cultural symbols and traditions" (1983, 13–14). Ultimately, however, historical communities emerge, organized by teachers and leaders who control the process of interpretation. "In the process the controlling group marginalizes and suppresses other branches of the community, with their own texts and lines of interpretation. The winning group declares itself the privileged line of true (orthodox) interpretation. Thus a canon of Scripture is established" (14). The contemporary community, however, stands in a peculiar relation to the historical community: orthodoxy may become meaningless in relation to present human experience. One enters the hermeneutic circle when the contrast between one's experiences and the received tradition becomes evident. "If a symbol does not speak authentically to experience, it becomes dead or must be altered to provide a new meaning" (12–13).

So framed, Ruether's *New Woman, New Earth* (1975), *Sexism and God-talk* (1983), and her latest book, *Gaia and God* (1992), collectively build a creation story that provides an alternative to that of conservative, moderate, and even other liberal Christians (since most of them do not incorporate feminist criticism). Early on, Ruether (1975) argued that patriarchy and patriarchal religion were largely responsible for the domination and oppression of women and nature. The Industrial Revolution and Enlightenment more generally intensified this long-established project. "The nineteenth-century concept of 'progress' materialized the Judeo-Christian God concept. Males, identifying their egos with transcendent 'spirit,' made technology the project of progressive incarnation of transcendent 'spirit' into 'nature.' The eschatological project [of historical Christianity] became a historical project [of the industrial growth society]" (194). The result is a culture that ruthlessly exploits nature and, ironically, has become godless. "With the secularization of society, religion and morality become feminized or privatized. Morality becomes appropriate only to the individual person-to-person relation exemplified by marriage. . . . A morality defined as 'feminine' has no place in the 'real world' of competitive male egoism and technological rationality" (199).

According to Ruether, feminist theology is a systematic correction "of each category of Christianity," a hermeneutic that restates and transforms the Christian paradigm (1983, 38). Traditional stories, on her interpretation, are inhabited by a Christian metaphysics of presence—an androcentric bias that creates *an ontological divide between spirit,* as epitomized by God and Man, *and matter,* as epitomized by Nature and Woman. Nature religions (as in Goddess-worshiping cultures) did not divide spirit and matter, heaven and earth, but conceived of them "as dialectical components within the primal matrix of being. . . . Spirituality was built on the cyclical ecology of nature, of death and rebirth. Patriarchal religion split apart the dialectical unities of mother religion into absolute dualism, elevating a male-identified consciousness to transcendent authority" (1975, 194–95).

Ruether reads the ancient Jewish community as having likely edged the West toward dualism. "Whereas ancient myth had seen the Gods and Goddesses as within the matrix of one physical-spiritual reality, male monotheism begins to split reality into a dualism of transcendent Spirit (mind, ego) and inferior and dependent physical nature" (1983, 54). Articulated through time, this initial division becomes the Great Chain of Being, with God the Father over all. God's creation is Man, as exemplified by Adam, who is made in his image, and Nature. Woman comes from Adam, God's first human creation. Not only does woman become secondary in relation to God, but woman also has only a negative rather than an affirmative definition. "Whereas the male is seen essentially as the image of the male transcendent ego or God, woman is seen as the image of the lower, material nature. Although both are seen as 'mixed natures,' the male identity points 'above' and the female 'below.' Gender becomes a primary symbol for the dualism of transcendence and immanence, spirit and matter" (54).

By rereading the biblical tradition, Ruether reestablishes connections for Christians with the earth and its flora and fauna. She argues that the "Big Lie" of modernity—that human beings are above nature and therefore naturally dominate it—is giving way to "Divine Wisdom." The Big Lie is rooted in the Christian creation story, which is built on a hierarchical model "that starts with nonmaterial spirit (God) as the source of the chain of being and continues down to nonspiritual 'matter' as the bottom of the chain of being and the most inferior, valueless, and dominated point in the chain of command" (1983, 85). Divine Wisdom leads to a new creation story that no longer abstracts God from the cosmos and humankind from the earth. In effect, feminist narrative prepares the way for an alternative creation story. "The God/ess who is primal Matrix, the ground of being-new being, is neither stifling immanence nor rootless transcendence. Spirit and matter are not dichotomized but are the inside and outside of the same thing" (85).

The kenosis of the Father (*kenosis* means an emptying of the form of God) thus opens the door for "the Shalom of the Holy; the disclosure of the gracious *Shekinah*; Divine Wisdom; the empowering Matrix; She, in whom we live and move and have our being—She comes; She is here" (1983, 266). From Ruether's perspective such a creation story helps humankind begin to care for the earth: it transforms our relations to nonhuman others from one of domination and exploitation, legitimated by patriarchal religion, to one of liberation and nurture. We begin to "respond to a 'thou-ness' in all beings. . . . We respond not just as 'I to it,' but as 'I to thou,' to the spirit, the life energy that lies in every being in its own form of existence" (87). Such a new creation story helps to transform culture: from managing nature to minding nature. In Ruether's view minding nature means grasping the logic of ecological relationships and fitting human ecology together with the ecology of the more than human in ways that optimize, rather than undercut, outcomes for the entirety of the Creation.

Gaia and God, Ruether's latest work, extends her project, offering extended analyses of classical creation stories and religious narratives of world destruction, as well as explorations of possibilities for healing the earth and ourselves. The title of the book is a deliberate juxtaposition of terms, designed to "pose the question of the relationship between the living planet, earth, and the concept of God as it has been shaped in the Western religious tradition" (1992, 4). Ruether's analysis sharpens the critique of patriarchal religion with its distorting consequences for Christians and begins to develop an explicitly stated ecological ethic. She argues that much of the problem with Christianity is its two inconsistent, even contradictory positions toward the Creation. "On the one hand, humans are said to be guilty for the inadequacies of the rest of nature [through original sin]. . . . On the other hand, humans bear no ultimate responsibility for the rest of creation. Animal and plant life can be exploited at will by humans as our possessions. They have no personhood of their own that need be respected, and we share no common fate with them" (30).

The solution for Christians lies in creating an ecological ethic that reintegrates Gaia and God, that puts together what patriarchal religion and materialism (part of the patriarchal project) have put asunder. "An ecological ethic must be based on acceptance of both sides of this dilemma of humanness, both the way we represent the growing edge of what is 'not yet' of greater awareness and benignity, and also our organic mortality, which we share with the plants and animals" (1992, 31). We humans are, Ruether insists, embedded in the Creation, the natural world. We work out our Christian hopes and dreams, our aspirations, in this world, "partly by shaping 'nature' to reflect these human ideals. But this reshaping is finally governed by the finite limits of the interdependence of all life in the living system that is Gaia" (31).

In weaving an ecotheology, Ruether does not force Christians to choose Gaia over God, or God over a concern for Gaia. Rather, she calls for a subtle reorientation that embraces both. The thunderous, masculine voice of God has drowned out the softer, more feminine voice of the earth. "Her voice does not translate into laws or intellectual knowledge, but beckons us into communion" (1992, 254). But communion alone is not enough. "We need organized systems and norms of ecological relations" (255). And Christians need to "revisit the questions of good and evil, sin and fallenness," and other relevant aspects of the Judeo-Christian tradition. "We inherit in our Christian tradition (as do others in their traditions) both cultures of domination and deceit and cultures of critique and compassion" (258). It is the latter aspect of Christianity in which Ruether places her faith for creating and politically empowering an ecological ethic.

Judaism

Liberal Jewish creation stories, unlike those of Protestants or Catholics, do not involve a Christology. But like other liberals, reformed Jews employ science-based narratives in reweaving their creation stories. For the liberal Jew the divine (Yahweh) is still revealing itself in the ongoing processes of the world. Martin Buber's writings offer an example of a liberal Jewish theology that could reinforce an ecocentric ethic generally and a metaphor of caring for creation specifically.

The anthropocentrism of the modern age, as Buber has it, is a consequence of the status we assign to "subject-object" ways of thinking, ways of thinking that privilege human beings as thinking subjects over natural entities and processes as mere objects. This way of being-in-the-world is clearly a historical artifact, the consequence of the narratives of the scientific revolution. The aftermath is that the modern person lives in, as Buber terms it, an *I-It world*.

Not only are our relations to nature primarily objective, but we exist socially as objects externally related, through the market, to one another. Every individual has been conditioned to interpret experience in terms of abstract conceptual schemes, such as scientific or economic theory. Accordingly, we view the earth as nothing more than a stage on which we enact the human drama. We view flora and fauna as mere natural resources to fuel the human project. Buber argues that the basic relation to the It-world is one of utilization. "What has become an It is then taken as an It, experienced and used as an It, employed along with other things for the project of finding one's way in the world, and eventually for the project of 'conquering' the world" ([1923] 1970, 91).

Buber does not argue that humans need to abandon objectivity completely. The problem with the I-It relation to the world is that it obscures the primordial relation I-*You*, a relation that is fundamental to all being. What is crucial, he insists, is the realization that *every thing*, including not only other human beings but also the earth and its flora and fauna, exist in a potential I-You relation. Obviously, an I-You relation to animals and plants is not the same as a similar relation to a person or to God. Buber distinguishes among three spheres of I-You relations. "The first: life with nature. Here the relation vibrates in the dark and remains below language. The creatures stir across from us, but they are unable to come to us, and the You we say to them sticks to the threshold of language" ([1923] 1970, 56–57). Human relations constitute a second mode of being in which we enter the domain of language. "We can give and receive the You" (57). Finally, human beings have relations with spiritual beings (with God). "Here the relation is wrapped in a cloud but reveals itself, it lacks but creates language" (57). I-You relations with the inorganic world are "sub-threshold," meaning that there cannot be complete mutuality as with human beings; relations with the infrahuman world of plants and animals are "pre-threshold." Nonetheless, natural objects and agents can be encountered and appreciated as any You—that is, in their otherness, in their being-for-themselves rather than in their being-for-humans.

I-You encounters with the world are relational, being-to-being. As the knower "beholds what confronts him [or her], its being is disclosed. . . . It is no longer a thing among things or an event among events; it is present exclusively" ([1923] 1970, 90). Further, an I-You relation is not one of an individual entity being reduced to a mere instance of a universal but rather one of genuine individuality. There is a genuine exchange or presence of one to another one. With infrahuman species, however, this relation is outside language. Buber equates, for example, the gaze of an animal with language. But "this language is the stammering of nature under the initial grasp of spirit, before language yields to spirit's cosmic risk which we call man" (144–45). He also points out how tenuous and fleeting our genuine contacts with infrahuman others are, for the I-It relation quickly overrides them. He mentions a gaze shared with his cat, where for an instant "the bright You appeared and [then] vanished: had the burden of the It-world really been taken from the animal and me for the length of one glance? At least I could still remember it, while the animal had sunk again from its stammering glance into speechless anxiety, almost devoid of memory" (146).

Buber also discusses I-You transactions with the botanical world, as for example with a tree. Again, the relation between a person and a plant is not exactly the same as the relation between two human beings, since the tree, like the animal, lacks language. Further, the tree is not capable of responding as an animal (as with a gaze, a movement, or a sound). And like the animal, the I-It relation continually intrudes, for humans have been conditioned by language to thrust the tree into categories. For example, Buber writes, "I can assign it to a species and observe it as an instance with an eye to its construction and its way of life" ([1923] 1970, 57). Yet there is potential for something more, Buber argues, than an I-It relation. "It can also happen, if will and grace are joined, that as I contemplate the tree I am drawn into a relation, and the tree ceases to be an It. . . . Does the tree then have consciousness, similar to our own? I have no experience of that. . . . What I encounter is neither the soul of a tree nor a dryad, but the tree itself" (58–59).

However tenuous and fleeting it may be, a *dialogic relation* to nature is quite different than the I-It transactions of the everyday world. For nature manifests, through I-You relations, its intrinsic value, which precedes whatever extrinsic (utilitarian) value it has in the It-world. As Ray Barlow argues in his interpretation of Buber, an I-You encounter with the earth and its creatures transforms the objective knower into a "subjective person engaged in existential relation with an individual being, and finally relates to the pre-threshold being without ulterior, anthropocentric purposes. The very purposelessness of the encounter suggests that the human knower is allowing the pre-threshold to exist on its own terms, rather than ordaining it to human use" (1989, 19).

Buber also argues, in *I and Thou*, for a direct dialogue between God and the individual; in fact, he believes that the crux of biblical religion is the dialogue between human beings and God in which each becomes the other's You. Barlow argues that this dimension of Buber's thought makes possible a genuinely ecocentric reading of *I and Thou*. "The divine encounter therefore, is a perspective for all knowledge of the world, for a world-view that reflects the commonality of being, as well as a guarantor of the importance of each relation which comprises the *cosmos* and a basis for an objective explanation of the relation of human and natural beings" (1989, 24). If Barlow is correct, then Buber has come nearly full circle to the Old Testament creation story and the tradition of the covenant. For the Creation is not an It-world given to humankind to manipulate for pleasure, profit, or any other human scheme. "Solely in the relation to God," Buber argues, "are unconditional exclusiveness and unconditional inclusiveness one in which the universe is comprehended" ([1923] 1970, 148).

Radical Creation Stories

Radical creation stories go beyond, though they do not abandon, a Judeo-Christian frame of reference. Indeed, they characteristically challenge the religious mainstream to renew itself in a truer, deeper faith. Unlike liberals, who seek rapprochement between scientific and Judeo-Christian narratives, the relation is characteristically reversed, and Judeo-Christian traditions are set in the context of scientific discourse. The best known radical creation story is the *new cosmology*, an area of ecotheology in which the work of Thomas Berry stands out.[26] Berry, though he is a Passionist priest in good standing, cannot be construed as a liberal Catholic. He believes that Christianity has lost touch with its roots in a *creation mystique* and is now caught up in a *redemption mystique* that "is little concerned with the natural world. [For traditionalists] the essential thing is redemption out of the world through a personal savior relationship that transcends all such concerns" (1988, 129). Berry believes that it is best to let orthodox religion rest for now. One reason is the institutionalized dogma and character of the church. It is "difficult to find a theological seminary in this country that has an adequate program on creation. . . . The theological curriculum is dominated by a long list of courses on redemption and how it functions in aiding humans to transcend the world, all based on biblical texts" (133).

On the other hand, Berry does not believe that a strictly scientific creation story is adequate to a culture living in a time of ecocrisis either, since the focus is on a merely material process that does not include, in fact excludes, the spiritual. Our colleges and universities, he explains, do not prepare young men and women to live in a meaningful, coherent, integral world. "There is an inability to bring together the scientific secular world with the religious believing world or with the humanist cultural world. Each of these feels impelled to go its own way" (1988, 97–98). The consequence of religious discourse mired in a redemptive mystique and of science mired in physicalism, Berry argues, is the trivialization of both traditions. For Berry, religionists caught up in supernaturalistic dogma remain indifferent to the earth, and scientists caught up in naturalistic dogma remain blind to the spiritual aspects of the Creation. Thus the importance of the new cosmology.

Berry claims that the new cosmology enables an integral vision—a way to grasp the world, its creatures, and humankind as part of a purposive, cosmic continuum. The new cosmology is effectively a new revelatory experience (general revelation) that, as it articulates itself, is pushing us into "a new mythic age" (1988, 132). Teilhard, of course, strongly influences Berry, but Berry employs "the new science" of the late twentieth century—material that Teilhard did not have. The new science is a complicated subject, to be sure, but it can be characterized in part by the notion that the cosmos is self-creating, that our orderly universe has emerged from primal chaos.[27] For Berry the universe is "the primary revelation of that ultimate mystery whence all things emerge into being" (1987d, 107). Further, the cosmos seems to indicate the cogency of the weak anthropic cosmological principle: that there is no cogent account of our own existence (as carbon-based, big-brained bipeds) apart from a fifteen-billion-year evolutionary narrative. The human being, he writes, "is that being in whom the universe activates, reflects upon, and celebrates itself in conscious self-awareness" (108).

Berry suggests that *a fusion between religion and science* is occurring through the new creation story. It is fascinating, he writes, "to realize that the final achievement of our scientific inquiry into the structure and functioning of the universe as evolutionary process is much closer to the *narrative mode of explanation given in the Bible* than it is to the later, more philosophical mode of Christian explanation provided in our theologies" (1988, 136, my emphasis). This is an important distinction from Teilhard's creation story, which characterizes the evolutionary process as movement toward Point Omega. For Teilhard, Point Omega is metaphysical, something beyond natural process, a cosmic goal that exists outside of space-time. For Berry the new story escapes the "limitations of the redemption rhetoric and the scientific rhetoric" by achieving a new "integral language of being and value" (136). In short, the new story is a happening, something inside and not outside the cosmic continuum, a literal part of cosmic process.

For Berry the new creation story is a sacred canopy: a new legitimating narrative for a global culture in which all creation lives and moves and has being. "If the way of Western civilization and Western religion was once the way of election and differentiation from others and from the earth, the way now is the way of intimate communion with the larger human community and with the universe itself" (1988, 136–37). Berry believes (as do some liberals, such as Cobb) that Christian spirituality itself will play an important role in the evolutionary process that is moving us from a failed cultural story, which sunders nature and spirit, toward a new story that rejoins them. Using language rich with millennial metaphor, he suggests that "the time has come for the most significant change that Christian spirituality has yet experienced" (117). But this new age is unlike the traditional vision of an "age of peace, justice, and abundance to be infallibly attained in the unfolding of the redemptive order" (114). It is also unlike the modern materialistic vision of economic success as the summum bonum. Judeo-Christians must transcend any previous expression of their faith. "The main task of the immediate future," Berry explains, "is to assist in activating the inter-communion of all the living and non-living components of the earth community in what can be considered the emerging ecological period of earth development" (1987d, 108).

The Universe Story (1992), Berry's most recent book, is a collaborative effort with the physicist Brian Swimme, and it combines science and religion to reach what might be termed a "mythic vision," which Berry and Swimme believe is essential to preserving the integral functioning of the Creation. Science, in short, offers a new revelation of the divine. "To preserve the natural world as the primary revelation of the divine must be the basic concern of religion" (242). In this context, Berry and Swimme perceive humankind as fallen, for "we have moved from such evils as suicide, homocide, and genocide, to biocide and geocide, the killing of the life systems of the planet and the severe degradation if not the killing of the planet itself" (247). But there is hope and opportunity in the new mythos, the universe story, for humankind is "expected to enter into this process within those distinctive capacities for human understanding and appreciation that provide our human identity. We are expected to enter into the process, to honor the process, *to accept the process as a sacred context for existence and meaning*, not to violently seize upon the process or attempt to control it to the detriment of the process itself in its major modes of expression" (251–52, my emphasis).

Matthew Fox, who trained as a Dominican priest, has also developed an explicitly ecotheological, radical Christian position. In 1988, Fox was silenced for one year by his order (under pressure from the Vatican), and he has since been dismissed by the Dominicans. In part, Fox came under pressure by the

Catholic hierarchy for his "deep ecumenism," his belief that all faiths are revelations of the divine, and his avowal of a creation spirituality. A cosmic Christology figures prominently in Fox's theology; he equates the human exploitation and desecration of nature with the crucifixion and resurrection of Jesus. "Mother Earth is being crucified in our time and is deeply wounded. Like Jesus at Golgotha, she is innocent of any crime, 'like us in every way save sin' (Heb. 4:15). . . . Yet, like Jesus, she rises from her tomb every day" (1988, 145). Unlike Teilhard, whose Christology is strongly associated with a future Point Omega, Fox's Christ is Mother Earth, here and now. Mother Earth, he writes, "is the temple, the sacred precinct in which holy creation dwells and praises God. Jesus' mission and message were rejected in his lifetime—that is what his crucifixion symbolizes. That air of rejection is what makes our mistreatment of Mother Earth so poignant" (146).

Fox employs biblical source material to make his case. On his reading, Joel 1.1–12 is a lamentation over the ruin of the landscape. "Stand dismayed, you farmers, wail, you vine dressers, for the wheat, for the barley; the harvest of the field has been ruined. The vine has withered, the fig tree wilts away; pomegranate, and palm, and apple, every tree in the field is drooping. And gladness has faded among the sons and daughters of the human race." Further, Fox argues, Joel 1.13–2.17 is a call to redeem the country through repentance and prayer. "But now, now—it is Yahweh who speaks—come back to me with all your heart, fasting, weeping, mourning. Let your hearts be broken, not your garments torn, turn to Yahweh your God again, for Yahweh is all tenderness and compassion." This passage, Fox suggests, "corresponds to that awakening of heart knowledge and of entering the darkness that is called mysticism. . . . With this breakthrough the day of Yahweh—a day of compassion and justice, of healing and celebration—can truly happen" (1988, 4).

Fox's ecotheology weaves in narrative traditions, especially Christian mysticism, that conservative and even moderate Christians do not accept. In an argument that takes up the twelfth-century abbess Hildegard of Bingen's writings, he contends that envisioning Jesus as a symbol for the earth is appropriate, since the earth is holy both in itself and "because it provided the body by which the Son of God was made human flesh" (1988, 146). Following Paul's description of Jesus as "first-born" and as the "first-fruit" (Rom. 8.28; 1 Cor. 15.20, 23), Fox argues that these images are "part of the Jewish paschal imagery of fertility rites which celebrate Passover as a 're-enactment of the cosmogony or creation' " (147). And following the fourteenth-century German mystic Meister Eckhart's belief that "every creature is a word of God," Fox writes that wisdom has been made flesh not in Christ but in all his myriad expressions. "Mother Earth," he concludes, "is a special word of God: a unique expression of divine wisdom, of divine maternity and caring, of divine creativity and fruitfulness" (147).

Another kind of radical creation story is offered by *nature religionists*—a persistent American tradition that has understood nature as sacred. Katherine Albanese argues in *Nature Religion in America* that the continuity of this tradition "is one more sign that, in a 'secular' society, the search for the sacred refuses to go away" (1990, 201). Similarly, Mark Sagoff contends in *The Economy of the Earth* that in North America, even from colonial times, nature was historically a symbol of the divine; the wilderness assured the colonists "of their special relation to God" (1988, 133). Albanese lists a variety of traditions, including aboriginal (Native American), Goddess feminist, and transcendentalist strains. Some of these, such as Goddess feminism and aboriginal traditions, are best described in the following section. Here my focus is on John Muir, who found in the American wilderness a divine presence still at work in the act of creation.

Though raised in a conservative religious household, Muir came to reject any orthodox creation story as the work of a transcendent and eternally separate God. From his own observations of evolutionary process in the Sierra Nevadas he knew that "the world, though made, is yet being made; that this is still the morning of creation" ([1915] 1979, 67). For Muir nature is a visible manifestation of God. God is not apart from creation but within creation; in this manner the traditional God of Judeo-Christianity, separate from and above nature, becomes a God incarnate, part of an ongoing creation. Michael P. Cohen suggests that Muir came to read the creation as a book, the Book of Nature, and that he "read it largely as an early nineteenth-century scientist might, with the assurance that it was a sacred book" (1984, 109). Ultimately, Muir believed that God was nature, thereby serving the motive for metaphor—that is, the human desire to identify with the world of which we are a part. Muir is best read as a pantheist, as a nature religionist who found God and celebrated the divine presence in the wilderness.

For Muir the Creation is a natural community in which all creatures have a purpose (intrinsic value) and no species enjoys privileges. Further, and crucially, the Creation remains in process, a vital disclosure of God incarnate, a temporal manifestation of a divine soul. "What is 'higher,' what is 'lower' in Nature?" he asks ([1938] 1979, 137). And he answers, "All of these varied forms, high and low, are simply portions of God radiated from Him as a sun, and made terrestrial by the clothes they wear, and by the modifications of a corresponding kind in the God essence itself" (138). As Muir wandered the mountains of California he witnessed the evolutionary process at work. Traveling in Alaska, he came to believe that "in very foundational truth we had been in one of God's own temples and had seen Him and heard Him working and preaching like a man" ([1915] 1979, 68). This position divorces Muir from conventional Judeo-Christian theology (theism), for his God is humanlike—temporally integrated with a natural world. By making divinity incarnate, providence is denied. There can be no supernatural spectator apart from the world with a divine plan in mind; and humankind can no longer be the chosen species awaiting eternal salvation while witnessing an essentially meaningless because preordained passage of time. Human beings are bound with time, in process, like the entire Creation. And after we have played our part, Muir writes, our species, too, "may disappear without any general burning or extraordinary commotion whatever" (1916, 140).

For nature religionists like Muir the creator does not privilege human beings. Humankind is not the apex of God's work, and human purpose is no more important than any other. "There are," he argues, "no square-edged inflexible lines in Nature. We seek to establish a narrow line between ourselves and the feathery zeros we dare to call angels, but ask a partition barrier of infinite width to show the rest of creation its proper place" (1954, 313). Muir doubts that God gives the creation to "Lord Man" to do with as he sees fit. One reason is that those "profound expositors of God's intentions" who argue that the Creation serves *man alone* ignore the other side of nature, such as human-eating animals and crop-eating insects (1916, 138). It does not occur to Christians, he continues, that

> nature's object in making animals and plants might possibly be first of all the happiness of each one of them, not the creation of all for the happiness of one. Why should man value himself as more than a small part of the one great unit of creation? And what creature of all that the Lord has taken the pains to make is not essential to the completeness of that unit—the cosmos? The universe would be incomplete without man; but it would also be incomplete without the smallest transmicroscopic creature that dwells beyond our conceitful eyes and knowledge. [1916, 138–39]

Muir is arguably the most important American nature religionist insofar as nature religion places Judeo-Christianity in the context of evolutionary science. As Albanese points out, however, some nature religionists never go beyond orthodoxy. For the New England Transcendentalists, epitomized by Ralph Waldo Emerson, nature was not God but was symbolic of the Cosmic Creator, a material emblem for the spirit of God. Albanese also finds in nature religion a range of belief: from those who sought to live simply and piously in a harmonious relation with nature to those who believed that the harmony of nature allowed humankind to dominate the land and its flora and fauna. Some nature religionists, she argues, remained enframed within the outlines of conventional Judeo-Christian orthodoxy, believing that God had made the Creation to serve his children, fashioned in his image. Yet nature religionists, such as our contemporary David Douglas, make the obvious counterargument. If we believe that the earth is God's creation, then "we would act less recklessly . . . , not only in irreversible ecological affairs, but in quieter relationships with the earth and its creatures day to day" (1987, 30).

Alternative Creation Stories

By definition (see figure 2, above), alternative creation stories are outside the Judeo-Christian frame. Some, such as those of Native Americans, are indifferent to Judeo-Christian orthodoxy, while others, such as Goddess feminism, vigorously reject Judeo-Christianity, implicating it in the suppression of their traditions. There is some temptation to lump all alternative creation stories into the nature religion category, since they characteristically view nature as sacred. But there are also striking differences among alternative creation stories, especially between those that spring from feminist narratives and those traditionally held by Native Americans. Among alternative creation stories are those of postpatriarchal feminists, such as Goddess feminists, so-called radical feminists, and Wiccans, who (unlike Rosemary Ruether, the postpatriarchal Christian feminist discussed above) uniformly believe that Judeo-Christianity is so fundamentally flawed that women must establish their own spiritual movements.

Post–Judeo-Christian Radical Feminist Ecotheology

A useful point of reference in this faith tradition is the work of Mary Daly. In "Original Reintroduction" (1985), she argues that little has changed since her book *Beyond God the Father* was published in 1973. According to Daly, Judeo-Christians continue to reify God as the Father rather than seeing God as a verb. To think of God as a verb, she continues, "expresses an Other way of understanding ultimate/intimate reality. The experiences of many feminists continue to confirm the original intuition that Naming Be-ing as Verb is an essential leap in the cognitive/affective journey beyond patriarchal fixations" ([1973] 1985, xvii). This "Other way" for Daly is distinctively and even exclusively feminist. Clearly, then, her story repudiates Judeo-Christian narrative traditions, including the creation story. She invites women into a radical form of discourse. On her analysis Judeo-Christian narratives are words of domination or power, "ruts, already violently embedded into women's psyches, which track women into dying out their lives in patterns of pointless circling and re-acting" (xiv). Daly counsels women to resist the temptation to "re-turn" to Judeo-Christian traditions and instead to find their own words of power: Goddess-words and discourse. Goddess-words, she argues, "become our broom, our flying Night-mare, carrying Wild wanderers beyond the dulling daydreams programmed by the perpetual soap operas of the sado-state" (xix).

Daly's *Gyn/Ecology* can be read, she suggests, as a radical feminist creation story that attempts to break free of "all forms of pollution in phallo-technic society," that is, "the mind/body/spirit pollution inflicted through patriarchal myth and language on all levels" (1978, 9). The book's title is itself the beginning of a journey of creation for women, "a way of wrenching back some wordpower. The fact that most gynecologists are males is in itself a colossal comment on 'our' society" (9). The devastation of the natural world and of woman's bodies and psyches are, for Daly (and all ecofeminists), part of the same patriarchal complex. To escape that complex requires a new story. *Gyn/Ecology* at least sets that project in motion.

> That is, it is about dis-covering, de-veloping the complex web of living/loving relationship *of our own kind*. It is about women living, loving, creating our Selves, our cosmos. It *is* dis-possessing our Selves, enspiriting our Selves, hearing the call of the wild, naming our wisdom, spinning and weaving world tapestries out of genesis and demise. In contrast to gyn-ecology, which depends upon fixation and dismemberment, Gyn/Ecology affirms that everything is connected. [10–11]

Wiccans

Starhawk, the author of several books and many articles, is perhaps the best known practitioner of the so-called Old Religion. She writes, "We are called Witches, a word that stems from an Anglo-Saxon root meaning 'to bend or shape.' Witches were shamans—benders and shapers of reality—but shaman is a word so overused and commercialized today that I don't claim it for present-day Witches" (1989, 175). One of Starhawk's aims is to recover ancient traditions of nature worship. Another goal is to undo what she perceives as many centuries of propagandizing by Christians to equate the Old Religion with Satanism. Like Daly, Starhawk believes that feminism naturally leads to an earth-based spirituality, a spirituality that denies the separation between the natural and the divine. Wiccans believe sky-gods to be an invention of the patriarchal mind.

An earth-based spirituality, according to Starhawk, has three principles at its core. The first is the principle of divine immanence in the Creation. This concept leads naturally to a caring for creation rather than its domination and exploitation. "When the sacred is immanent, each being has a value that is inherent, that cannot be diminished, rated, or ranked, that does not have to be earned or granted" (1989, 177). The second principle is the concept of interconnection, comprising not only connections among the parts of the human body but links with the living body of the earth. "This deep connectedness with all things translates into compassion, our ability to feel with and identify with others—human beings, natural cycles and processes, animals, and plants" (178). Last is the principle of compassion. "Compassion allows us to identify powerlessness and the structures that perpetuate it as the root cause of famine, of overpopulation, of the callous destruction of the natural environment" (180). In aggregate these three principles speak to our "deepest experiences" of the world, Starhawk claims. "When you understand the universe as a living being, then the split between religion and science disappears because religion no longer becomes a set of dogmas and beliefs we have to accept even though they don't make any sense, and science is no longer restricted to a type of analysis that picks the world apart" (1990, 73). By emphasizing the importance of sacraments and rituals as well as the three principles, Starhawk can argue that an earth-based spirituality is more than an intellectual exercise (or traditional theology); it is a practice—a way of life. "For those of us called to this way, our rituals let us enact our visions, create islands of free space in which we can each be affirmed, valued for our inherent being. In ritual we can feel our interconnections with all levels of being, and mobilize our emotional energy and passion toward transformation and empowerment" (1989, 184).

Goddess Feminism

Elinor Gadon's *Once and Future Goddess* makes abundantly clear the concern of Goddess feminists for life on earth—all the flora and fauna as well as global ecological processes. Goddess feminists argue that the abuse and exploitation of nature that created ecocrisis is linked to the historical oppression and domination of woman. Although this thesis is controversial, a wealth of scholarship—such as Carolyn Merchant's *Death of Nature*, Gerda Lerner's *Creation of Patriarchy*, and Susan Griffin's *Woman and Nature*—that is not allied with but complementary to Goddess worship, adds cogency to it. Lerner, for example, argues that Hebrew monotheism mounted a relentless assault on Goddess traditions. The outcome was that the Genesis tradition replaced the female role in creation with an all-powerful male God. Female sexuality in turn was associated with evil and a fallen material order of natural being. "This symbolic devaluing of women in relation to the divine becomes one of the founding metaphors of Western civilization" (1986, 10).

Goddess feminists thus believe that Judeo-Christian creation stories must be abandoned. "The familiar story we know so well from the biblical account in Genesis is outdated," Gadon contends, "no longer serving us well when life as we know it on our planet is threatened with unprecedented ecological disaster" (1992, 187). Conventional wisdom, she continues, is that we live in a secular society based on an Enlightenment narrative tradition that separates church and state. The reality is that "the Judeo-Christian worldview continues to undergird Western culture. We continually find ourselves mired down in the miasma of time-worn mythology and symbols that define our identity and relationships. At some deep existential level our mythology is more real than scientific truths we claim to live by" (188). Part of the problem, Gadon believes, is the claim of Judeo-Christian orthodoxy that time is going somewhere: that is, the belief that we are the chosen people moving in historical sequence from the day of creation to the day of judgment. In adopting this belief, Judeo-Christians abandoned the primordial sense of time inherent in the Goddess tradition—the sense that time is cyclical and that human existence is bound up with the great cosmic round, especially the interweaving of life and death. "Linear time moves relentlessly towards death and the Day of Judgment in dramatic contrast to the ever-renewing cyclical time of worldviews that are not based on historic divine revelation. In the Christian orientation of Western culture, human sexuality is linked with death, rather than generativity, the source of the ongoing round of renewal; female sexuality and female bodies are considered evil" (183).

What the Goddess story offers, according to Gadon, is a way beyond our present suicidal impasse—where our life means the deaths of nature and ultimately our culture. Rather than the creatio ex nihilo of the Judeo-Christian creation story, which is archetypically male, the Goddess story recognizes the earth as creative. *The Once and Future Goddess* offers a remarkably detailed, cross-cultural study of historical Goddess cultures. These cultures revered the creative process, the sacred interweaving of life and death—as in the cycles by which plants nourished herbivores and herbivores, in turn, nourished humans. Crucially, however, Gadon does not believe that we late-moderns can simply return to the ancient ways of Paleolithic hunter-gatherers and Neolithic agriculturists, replacing our Judeo-Christian myths with theirs. But the Goddess myths of these ancient people inspire us in our own quest "for a way to resacralize our own relation to the earth" (1992, 196).

One way to achieve this is through metaphors of birthing, since birth has such obvious connections with creation as well as with processes of becoming and transformation. As Albanese points out, in using metaphor, Goddess feminism distinguishes itself from Judeo-Christianity, which typically revels in its literalness, in the truth value of its creation story (1990, 179). The metaphorical-symbolical aspect of Goddess feminism underlies its essential genius. The new iconography, Gadon argues, "is a key to our power as women to name the world as we see it" (1989, 372). But it does something else as well, offering a "bigger vision that includes both scientific thinking and the wisdom of the past" (374).

So conceived, the new iconography begins to counter the mechanistic materialism of modern science, which, operating historically in conjunction with the church, justified the domination and exploitation of nature: nature was mere atomic matter-in-motion. Today we know the earth is simply not the tidy system of scientific laws, precise measurements and equations, and theoretical models that Western culture assumes. The earth is far greater than the classical scientific worldview permits us to acknowledge, let alone imagine.[28] Further, the earth is not what modern economics represents it to be: standing reserve subject to human manipulation and appropriation. Seen from the vantage point of Goddess feminism, to destroy the earth is to murder the divine. Goddess worship does not, Gadon argues, mean a return to pagan religion. Rather, "the Goddess has reappeared . . . as a symbol of the healing that is necessary for our survival." In revering the Goddess we honor "all that lives—women, the earth, its manifold creatures—[and thus there is] no longer need to control, oppress, despoil our planet, to make war" (1989, 376).

Native American Faiths

Indigenous (American Indian) faith traditions also offer alternative creation stories that cannot be reconciled with Judeo-Christian narrative traditions.

While Judeo-Christians typically denigrate tribal religions as "barbaric and savage," Karl Luckert argues that they are better grasped as the surviving relics of "the oldest religion" of humankind, a religion that has been developing for some three million years (1975, 3). Judeo-Christianity is a relative newcomer, and its creation story perhaps the newest of stories. The creation stories of indigenous Americans are colored by a deep and abiding respect for all forms of life and the earth. Native peoples revered the land in a way that is difficult for Judeo-Christians to understand (although the tribes of Yahweh might have). Indeed, they might even be said to have identified with the earth, to have taken their sense of purpose from it through mimetic consciousness. Animals rather than other humans became role models, because the indigenous hunter-gatherers believed that they were morally and intellectually superior to humans (Spencer et al. 1977, 91).

Generalizations are subject to exception; in fact, Native Americans were never one single people, and they never had a single cultural pattern.[29] Some were arctic hunters; others were dry-land agriculturists for whom hunting was supplementary. As Albanese notes, every generalization runs the risk of doing "violence to the subjective sensibilities of many different peoples" (1990, 19). Yet these indigenous peoples also shared much. Albanese suggests that "Amerindian peoples lived symbolically with nature at center and boundaries. They understood the world as one that answered personally to their needs and words and, in turn, perceived themselves and their societies as part of a sacred landscape. With correspondence as a controlling metaphor, they sought their own versions of mastery and control through harmony in a universe of persons who were part of the natural world" (25).

In his book *God Is Red*, Vine Deloria, Jr., attempts a more comprehensive survey of Native American religion than Albanese. Deloria, recognized among his people and by scholars as an expert on tribal religions, argues that Judeo-Christian belief systems divorce human beings from any sense of a caring relation with the land. The ruin of the land is thus a consequence of Judeo-Christianity. In contrast, the traditional North American religions seek and recognize God (the holy, the sacred, the divine) in the landscape. For tribal religions, Deloria argues, creation is "an ecosystem present in a definable place," not as with Judeo-Christians "a specific event" (1973, 91). Further, American Indians do not believe that human beings were created in the image of God, although God is sometimes characterized as a grandfather. Rather, they identify with plants and animals. Their creation stories characteristically recall a primal flux, where humans lived in kinship with other animals and plants. All living beings are thought to have originally existed in this state of flux; any particular form, like human beingness, by which humans might be distinguished from, for example, wolves, is a matter of appearance rather than of ontological difference. After creation, forms became fixed, but all creatures remain kin, common beings of the world bound by familial obligation and respect into a cosmos—one cooperative whole.

The Zuñi tale "Creation and the Origin of Corn" catches this notion. Five things, as the story goes, are necessary "to the sustenance and comfort of the 'dark ones' [Indians] among the children of the earth. The sun, who is the Father of all. The earth, who is the Mother of men. The water, who is the Grandfather. The fire, who is the Grandmother. Our brothers and sisters the Corn, and seeds of growing things" (Cushing 1979, 346). So, too, the Navajo origin legend catches the sense of cosmic integrity in an ongoing creation.

> It is lovely, indeed, it is lovely indeed.
> I, I am the spirit within the earth . . .
> The feet of the earth are my feet . . .
> The legs of the earth are my legs . . .
> The bodily strength of the earth is my bodily strength . . .
> The thoughts of the earth are my thoughts . . .
> The voice of the earth is my voice . . .
> The feather of the earth belongs to me . . .
> All that surrounds the earth surrounds me . . .
> I, I am the sacred words of the earth . . .
> It is lovely, indeed, it is lovely indeed. [Iverson 1987, 56]

Deloria also suggests that a second major difference between Judeo-Christian and aboriginal creation stories lies in the characterization of the relation of humankind to nature. In the Judeo-Christian story, humankind enjoys dominion over nature, as in the naming of the animals and in the concept of stewardship. Deloria argues that the concept of stewardship is more part of the problem than any solution to ecocrisis. Such an attitude is "inadequate," he contends, "because it has not reached any fundamental problems; it is only a patch and paste job over a serious theological problem" (1973, 98). Aborigines, consistent with their creation story, are more concerned with the ongoing process of creation and the relations between things that comprise the Creation. Native Americans, it might be said, are more humble in their conception of the nonhuman world, believing not that they are above and superior to it but rather that they are a part of it. For Indians "the meaning of life comes from observing how the various living things appear to mesh to provide a whole tapestry" (102).

Yet another difference is the belief that all of the creation is alive: nature is not matter-in-mechanical-motion. Other living creatures, including plants, are alive; "they are peoples" just as other Indians are people (Deloria 1973, 103). For some native religions the idea that nature is alive extends even to inanimate entities, such as rocks. As Luckert argues, the animistic view of the cosmos characteristic of American Indians challenges Western rationalism and religion. But the apparent conflict between these two worldviews, he believes, reminds us of our choice. We must either choose once again to know the divine and to live in a sacred landscape or resign ourselves to an increasingly robotic existence inside the modern paradigm—the "World Machine" (1975, 5).

Deep Ecology

Last, ecophilosophy sometimes includes a religious dimension, and *deep ecology* is the most prominent example of this tradition. Arne Naess (1989), often recognized as the world's preeminent deep ecologist, makes clear that the path toward an ethical, caring orientation to the Creation can depart from a number of starting points, including Judeo-Christian, Buddhist, and Native American traditions (1989, 226). The work of Dolores LaChapelle is illustrative. In her many books and articles LaChapelle emphasizes a deep ecological consciousness that reconnects human beings with the nurturing earth. Her work resonates with the religious traditions of the native peoples of North America (and elsewhere)—a people who unlike Euro-Americans retain a sense of the land and its "people," that is, the flora and fauna, as sacred. Native Americans and other aborigines have three things, she argues, that Westerners lack: "an intimate, conscious relationship with their place," a sustainable economy, and "a rich ceremonial and ritual life" (1985, 247). LaChapelle points out that even a few hundred years ago Western culture still had seasonal festivals and rituals that celebrated the sacredness of the Creation. The purpose of these festivals, as implicit in the root meaning of religion (*religiare*, to relink), was to reconnect humans with the sacred landscape, to revive the *topocosm* ("*topo* for place and *cosmos* for world order"), that is, "the world order of a particular place" (248). The modern West, of course, no longer practices these rituals and seasonal festivals, and as a result we have lost the accompanying sense of reverence for creation.

Interestingly, LaChapelle traces this loss not to Judeo-Christianity but to the appropriation of Christianity by the Roman Imperium after Constantine.

The surest method to gain political control over conquered peoples, the Romans discovered, was to destroy their local gods and religious traditions. According to LaChapelle and others, including Elaine Pagels (1988), Augustine was quick to seize the advantage of cooperation with the state; when Catholicism became the state religion, its victory over Gnosticism was assured. The characteristic modern fear of nature and belief that the wilderness is a place of abomination are the reverberations in our minds of these now forgotten events. "With every tree or rock or spring suspect as the abode of a devil," LaChapelle writes, "and with death hanging over the heads of those who went to them for spiritual refreshment, it is no wonder that all wild places became feared and avoided" (1978, 45–46).

But the teachings of Jesus, LaChapelle argues, had nothing to do with this. Indeed, following the lead of Josef Pieper's *In Tune with the World* (1965), she argues that some Christian rituals themselves trace their roots back to pre-Christian nature rituals. Many Christian feast days occur on dates associated since time immemorial with nature festivals, especially those that mark seasonal changes. The winter solstice (December 22), for example, is the Christian feast day known as Christmas (December 25), and the summer solstice (June 21) falls on the Vigil of the Feast of John the Baptist (1978, 139). LaChapelle's *Earth Festivals* (1973) is in part an attempt to recover some of these ancient traditions. The book offers a variety of seasonal festivals and daily rituals that all people (including Judeo-Christians) can celebrate and practice to rekindle a sense of bonding to the earth. Rituals help us, she contends, escape the grip of an exclusively objective and therefore an anthropocentric and exploitative relation to nature. "During rituals we have the experience, unique in our culture, of neither *opposing* nature nor *trying* to be in communion with nature, but of *finding* ourselves within nature" (1985, 250).

The poetry of Gary Snyder, who is sometimes called the poet laureate of deep ecology, is also exemplary, combining aboriginal traditions with ecosophy (deep ecological wisdom) and Eastern religion (especially Zen) into what Snyder calls an old-new way. My discussion here is merely illustrative of the space Snyder opens to care for creation. The title of his best-known work, Pulitzer-prize-winning *Turtle Island* (1974), symbolizes, and the poetry therein recreates, an ancient sense of a living and sacred creation that is more ongoing process than one-time event. In many aboriginal traditions the cosmic turtle, resting in the sea, is the home of all creatures and plants, and they inhabit the turtle's carapace. In this mythopoesy the creation is literally alive with creatures and plants, not mere matter-energy moving in timeless patterns of motion established by a transcendent creator. Such a belief enables humans to feel at home in the world of rocks and things, bound together in the dance of life with a sense of the sacred that is here and now.

For Snyder the ancient ones, such as the Anasazi of the American Southwest, become exemplars. He does not think that we can go back, yet he believes that they can inspire us. For they were rooted in the earth in ways that elude modern Americans; steel and concrete have replaced sandstone, supermarkets and fast food have supplanted fields of corn and beans. The ancient ones were in touch with the divine, "sinking deeper and deeper in earth / up to your hips in Gods" (1974, 3). Their cosmos was ordered: the kiva—the holy lodge or sacred space, an axis mundi—central to existence. The feathers of the eagle, itself an instantiation of divinity, adorned the heads of the shaman, who ritualistically (using pollen to trace intricate designs that symbolized the structure of the cosmos) maintained the cosmic order. Nature was conceived as feminine and nurturing, and the people were bound with it in organic cycles, as in "women / birthing / at the foot of ladders in the dark" (3).

In Snyder's opinion, American society, which values the land only as an economic commodity, is really the exception, since most cultures have thought of the land as sacred. In the West, sacred space has been associated with the transcendental rather than the existential. As Snyder notes, the linking of religion to the land "remains virtually incomprehensible to Euro-Americans. Indeed it might: if even some small bits of land are considered sacred, then they are forever not for sale and *not for taxing*. This is a deep threat to the assumptions of endless expansive materialist economy" (1990, 81). Of course, Americans characteristically dismiss this sense of land as sacred as mere primitivism, and they hold up their own religious convictions as a yardstick by which to measure aboriginal beliefs. Yet as Mircea Eliade argues, the behavior of prehistoric and archaic peoples "forms part of the general behavior of mankind and hence is of concern to philosophical anthropology, to phenomenology, to psychology," and more generally to all the human sciences and humanities (1959, 15). The turtle island cosmogony, if nothing else, divulges a sense of the land that is concealed by our modern story.

The idea of the earth as one creation in ongoing process also opens up a sense that the food chain is sacramental rather than merely economic. Snyder, like LaChapelle, emphasizes the importance of ritual, especially in maintaining a sense of the web of life as sacred. For the modern person meat is a commodity, a product of a highly mechanized process of food production directed by powerful, for-profit corporations that yields an unending stream of reasonably priced, shrink-wrapped food—devoid of any sign of blood and any sense that a living creature died for it. For the aborigine, however, meat was "the flesh" of the sacred game. By partaking of this flesh in the prescribed way, aboriginal hunter-gatherers confirmed and maintained the order of creation. Unlike modern consumers, who are oblivious to the profound ethical implications inherent in their interactions with the more than human, the ancient ones acknowledged a more than economic relation to "the others"—to the rest of creation. In eating the sacred game, then, human beings were killing part of the divine. The ritualistic piling of bones or other remains and other ceremonies were conducted to preserve the cosmic order and to ensure the return of the animals. Other rites governed how game was stalked and killed, and the kill was often consummated by the ritualistic consumption of the animal's heart or blood. The hunter was thought to take on the animal's attributes—to become one with creation, however mystical and tenuous such a union seems to the modern mind. Joseph Campbell points out (1988) that many of these practices are mirrored in Judeo-Christianity, as in the communion.

Snyder argues that Euro-Americans, deeply mired in ecocrisis, could find the way to an ecologically sustainable and spiritually satisfying way of life by recapturing some sense of land as sacred. For some modern Americans this begins with wilderness pilgrimages, which Snyder describes as literally inspiring. The pilgrim's "step-by-step breath-by-breath walk up a trail, into those snowfields, carrying all on the back, is so ancient a set of gestures as to bring a profound sense of body-mind joy" (1990, 94). That is, it awakens a religious sensibility, a sense of "that which helps take us (not only human beings) out of our little selves into the whole mountains-and-rivers mandala universe" (94). As cultural beings, humans inevitably return to society, and it is this return to society, now inspired by a sense of the sacred, that is the heart of Snyder's message. The point of the pilgrim's "studies and hikes is to be able to come back to the lowlands and see all the land about us, agricultural, suburban, urban, as part of the same territory—never totally ruined, never completely unnatural. It can be restored, and humans could live in considerable numbers on much of it" (94).

Caring for Creation and Solidarity

"There is no society without language," suggests Julia Kristeva, "any more than there is society without communication. All language that is produced is produced to be communicated in social exchange" ([1981] 1989, 7). Utilitarian individualism creates, however, the illusion that humans are the masters and possessors of language. The truth is that we are embedded in language, as much made by it as making it. By comprehending ourselves as storytelling culture-dwellers we can recognize the common themes that sweep across Western culture. Embedded in our Judeo-Christian narrative tradition is the defining characteristic of Western civilization: the belief that *time is meaningful*, not sound and fury signifying nothing. And, as the metaphor of caring for creation makes clear, a second commonality is inherent in the Great Code: the belief that *the Creation is good*.

Religious discourse, whatever appearances might be, remains relevant to the here and now. A sociolinguistic perspective reveals the utility of religious discourse. As Kristeva suggests, the study of language helps "to clarify the unconscious mechanism of the social functions themselves." By attending to the sacred canopy created by religious narrative, we place ourselves "face to face with all the forgotten and censored things that enabled" the modern world to be created. Through the linguistic turn "our culture is being forced to question once again its own philosophical mastery" (1989, 329). From a sociolinguistic perspective we can also see that a new synthesis is possible. Through the metaphor of caring for creation people of faith can reach common ground, even though the roads taken to that consensus may differ.

Northrop Frye offers a useful way of grasping the metaphor of caring for creation, especially insofar as this metaphor remains associated with the God of Judeo-Christianity. In his reading of Exodus 3.14, in which God says, "I am that I am," Frye suggests that we try thinking of the word *God* as a "verb implying a process accomplishing itself" rather than categorizing the word *God* as a noun (1981, 17). Modern people like to think of God as a noun, on the analogy of a thing, a physical body, but the ancient Hebrews and early Christians probably thought of God as a verb. To think of God as a verb, Frye continues, involves thinking "our way back to a conception of language in which words were words of power, conveying primarily the sense of forces and energies rather than analogues of physical bodies" (17). Although traditionalists and even some moderates may have difficulty with Frye's thesis, they should bear in mind that he seeks to reinvest religious stories with a social potency they have lost in the modern age. If we follow Frye, we come to the startling realization that Nietzsche is wrong. God is not dead but "entombed in a dead language" (18). The notion of God specifically and religion more generally as a verb is developed in chapter 6.

So comprehended, *caring for creation becomes an activity* carried on in a variety of ways not only across the Judeo-Christian spectrum but within alternative faith traditions as well. By telling cosmic creation stories that again give human beings a home in nature, we might begin to enter a new phase of history. As has become abundantly clear to virtually all informed citizens, unless we reform the institutions and norms that control our behaviors (directing them in ecologically pathological ways), our actions will undercut the very meaningfulness of our lives in time. Given the foregoing spectrum of belief that converges on a metaphor of caring for creation, a startling new possibility for a politically efficacious consensus on an environmental agenda appears before us. As Brian Swimme argues, *the telling of creation stories is perhaps the most important political and economic act of our time* (1988, 47).

Obviously, the creation stories explored here are discordant in some ways. This diversity, rather than helping us come together, threatens to pull us apart. But the stories converge on caring for creation. This convergence is more important than any difference, since it portends the transformation of postindustrial society. Stephen Toulmin argues that there is likely more of a convergence between creation stories than a divergence, even between scientifically based and religiously inspired narratives (1982, 268–69). A specifically religious approach to cosmology, he suggests, "means resolving among other things to deal with, and relate to, all created things in ways appropriate to their relations to God" (269). But, he continues, this is not in principle any different from a cosmological narrative that situates human beings in some overall scheme of things that makes no reference to God. The issue turns, fundamentally, on some sense of a context that constrains and thereby guides human behavior. A creation story provides exactly such a context.

Toulmin remains agnostic on the issue of how far ecology can take us along the road to a theology of nature. The "history of strained relations between scientific theory and fundamentalist theology" give us reason to wonder (1982, 274). But the future is now, and there is sufficient convergence among science- and religion-based creation stories to move ahead. The argument is essentially this: on the specific issues that confront the industrial growth society, such as biodiversity or holes in the ozone layer, there will be little practical difference between a theocentric creation story and an ecocentric creation story because both frame human action in a larger (limiting) scheme of things. At this juncture a fully ecumenical theology of nature is not necessary, for there is ample ground for solidarity in the caring for creation metaphor alone.

Our assessment of the church's cultural mission depends in part on how well we think a society such as ours can function without its traditional heritage. . . . With the loss of the knowledge of the Bible, public discourse is impoverished. We no longer have a language in which, for example, national goals (that is, questions of meaning, purpose, and destiny) can be articulated. We try to deal with apocalyptic threats of atomic and ecological disaster in the thin and feeble idioms of utilitarianism or therapeutic welfare.—George A. Lindbeck, "The Church's Mission to a Postmodern Culture"

5 The Role of the Church

In his preface to *Religion and Environmental Crisis*, Eugene Hargrove argues that too much energy has been spent either blaming religion for environmental crisis or arguing for one religion as the right solution. His proposal that we find ways for religion to respond is all the more cogent, then, in the context of regional ecological dysfunction and potential global ecocatastrophe. One reason that amelioration of environmental crisis eludes us is that religion has a fundamental role to play in environmentalism, a function that nothing else can fulfill. Yet orthodoxy, both secular and religious, precludes recognition of this function.

A sustainable society must remain an unrealized dream unless that goal is empowered politically. Faithful people can care for creation; they can find justification within their religion for creating an environmental ethic and for taking appropriate action. A genuine commitment to care for creation might change the face of the political landscape, especially by empowering *citizen democracy*.[1] The church is the ideal place for the faithful to begin to exercise their power as citizens. Francis Moore Lappé argues that "effective use of power in public life means ongoing reflection on our actions and those of others" (1990, 19). But reflection, in her opinion and mine, draws on and builds from collective memory, from a shared narrative tradition. Habits of the heart or collective memory, according to Lappé, are essential "if we are to incorporate into our everyday living attention to the well-being of the larger environment—the biotic community of which our species is but one member. . . . As we consciously rekindle memory, we can better envision the future, taking responsibility for the impact of our choices on nonhuman life and the lives of generations to come" (20).

Political activism follows naturally for any community that grasps the responsibility to care for creation. Politics for the church community is not about power per se but about affirming and realizing the ethically good, the goals that affirm the Creation. Religiously inspired political activism does not entail the idea of a state religion: the separation of church and state is a legal convention that precludes it. But politically, that is, in terms of the give and take of democratic debate and opinion shaping, the separation does not exist, since religion legally and demonstrably influences the political process—elections, policy decisions, legislation, and even the administration of public policy. American history confirms that the biblical tradition has been a constant and vital public influence—even after the Revolution and the creation of a confederation institutionalized the separation of church and state. Religion figured prominently, as has been mentioned, in the nineteenth-century abolition of slavery and the twentieth-century civil rights movement, only two of many examples sweeping a moral spectrum from declarations of war to the prohibition of alcohol consumption.

Since religious discourse countervails the first language of utilitarian individualism, it is necessarily involved in resolving ecocrisis insofar as there is any probability (as distinct from possibility) of altering the prevailing economic conception of the common good *and* the institutional complex that manages the industrial growth society. The idea of a moral life, Alasdair MacIntyre argues, requires a practice by human beings who live it out. Narrative is essential to moral practice, he continues, because the

> grounds for the authority of laws and virtues . . . can only be discovered by entering into those relationships which constitute communities whose central bond is a shared vision of and understanding of goods. To cut oneself off from shared activity in which one has initially to learn obediently as an apprentice learns, to isolate oneself from the communities which find their point and purpose in such activities, will be to debar oneself from finding any good outside of oneself. [1984, 258]

The church is the one place where most Americans can find narrative traditions that provide an alternative to the language of expressive individualism. Narratives have, MacIntyre argues, "both an unpredictable and a partially teleological character. If the narrative of our individual and social lives is to continue intelligibly . . . it is always both the case that there are constraints on how the story can continue *and* that within those constraints there are indefinitely many ways that it can continue" (216). Clearly, believers might find the grounds through religious narrative to make the ethical commitments and begin to implement the practices that care for creation.

Such an idea threatens numerous expert constituencies that presently dominate the public conversation about ecocrisis, both the technocratic elite that aims to manage planet Earth and the philosophical elite that offers master theories. Environmental ethicists, I have argued, like to claim that any solution for ecocrisis involves the adoption of a new theory. Whatever the utility of ethical theory for environmental ethicists in their own professional (and very small) community, the essential sterility of that argument is confirmed by the paradox of environmentalism. MacIntyre argues that post-Enlightenment ethical discourse is a failed project (as Nietzsche announced in the nineteenth century).[2] "The most striking feature of contemporary moral utterances is that so much of it is used to express disagreements; and the most striking feature of the debates in which these disagreements are expressed is their interminable character" (1984, 6).[3] Which is to say that the ethical theory of the professional environmental ethics community is powerless to overcome the pervasive influence of utilitarian individualism, an ideology institutionalized in political and economic institutions. Further, ecophilosophical discourse offers its ethical insights and ecological panaceas in a language inaccessible to lay publics.

Kenneth Sayre notes that ethical theoreticians appear incapable of moving a democratic majority to support policies that will lead toward sustainability, arguing that "if norms encouraging conservation and proscribing pollution were actually in force in industrial society, it would not be the result of ethical theory; and the fact that currently they are not in force is not alleviated by any amount of adroit ethical reasoning" (1991, 200).

Liberals mount a more serious challenge to my thesis that the church is essential to action on ecocrisis. They claim that the separation of church and state is absolute, that the line of separation cannot be crossed in any way without violating constitutional guarantees. Perhaps, as proponents of citizen democracy argue, liberals are following a Lockean notion that the state exists only to protect property rather than the idea that the state, whatever its relation to the market, is also a means to express public values (citizen preferences) that go beyond private economic interests (consumer preferences). For "doctrinaire liberals" public values are merely the sum of private interests; the idea that public values might evolve through religious discourse that converges on a common center of caring for creation is a fiction. So, too, is the notion that participation of the church in public matters can strengthen rather than endanger democratic life. Liberals perceive religious groups as taking uncompromising stands on issues in ways that polarize the body politic rather than lead to solidarity. Further, doctrinaire liberals fear religionist majorities that might overwhelm secularists, politically forcing their will (outlawing abortion, promulgating creationist textbooks, and so on) on a defeated minority. The Enlightenment narrative of political rights, grounded in the metaphysics of individualism, is threatened by the church. According to this tradition, the good society is the aggregate of the goods of individuals. Rights mean essentially the freedom of action to pursue *individual visions* of the good life. Common allegiance is owed only to the "state religion," that is, the constitutional guarantee of individual rights.

Fearing the intrusion of private interests on public purpose, classical liberals ignore the diversity of opinion within religion. As Steven Tipton puts it, "The public church has almost never spoken with a single voice" (Bellah et al. 1990, 181). Indeed, the diversity of faith in the United States is one reason why the "gospel of greed" is our cultural lingua franca. The faithful profess beliefs that appear similar, such as the belief in God, yet the God of conservatives and the deity of the new cosmology, let alone that of the Goddess feminists, are enormously different. Believers seldom appear able to achieve solidarity on public issues. Liberals also ignore the fact that the faithful themselves are not immune to the siren song of individualism. The biblical tradition calls them to give their allegiance to God, and yet the influence of *expressive individualism* is evident in the ease with which Americans abandon their religious commitments in the quest for "meaningful experience."

Church and State Reconsidered

The orthodox view that church and state are separate spheres is one reason why religion has not effectively addressed environmental crisis. Yet the notion that the church is one thing, concerned with otherworldly, supernatural affairs, and that the state is something else, concerned with this world, our everyday economic and political affairs, cannot be sustained. A closer examination of the role of religion in personal life and in democratic society reveals a different picture. The genius of democracy is not that religion is precluded from influencing public affairs; rather, it is that *no single religion or group of religions is permitted to monopolize the state.* As Garry Wills observes, America was and remains an experiment, since no other Western society has launched itself without theological sanction. The constitutional separation of church and state "gave to religion an initial, if minimal, freedom from crippling forms of cooperation with the state. That, more than anything else, made the United States a new thing on the earth, setting new tasks for religion, offering it new opportunities" (1990, 383). This freedom has been and remains both an opportunity and a danger. Freedom offers religion the chance to find its own essence, to be unencumbered by the demands of politics and economics. Autonomy from the state also poses the threat that the church might become irrelevant in a nation whose legitimating purpose is increasingly economic—that is, the maximization of the GNP regardless of ecological and social consequences.

One thing is clear: from a historical perspective the role of religion in American society has been and remains in flux. The colonists who came to the New World believed in religious establishment, as the history of the Massachusetts Bay Colony certainly shows. Even after the Revolution, establishment continued in some states (the First Amendment precluded establishment at the federal but not the state level). Massachusetts did not give up establishment until 1833. Still, the privatization of religion is one thing. The influence of religion and the church on society is another. Observing the new democracy during the 1830s, Tocqueville concluded that the church, especially by shaping the personality and moral character of individuals, was the primary political institution in the United States. Tocqueville's view, Bellah suggests, is that the political function of religion "was not direct intervention but support of the mores that make democracy possible. In particular, it had the role of placing limits on utilitarian individualism, hedging in self-interest with a proper concern for others" (1985, 223).

Clearly, even with the privatization of religion, the church exerts considerable political influence. Many of the faithful recognize that organized religion has an important function to serve in the public realm. Commentators like Wills, Tipton, and Marty argue that the role of the church is at least in part to criticize the state, sometimes even to castigate and break its rules, as in acts of religiously motivated civil disobedience. As discussed earlier, the opposition of church and state has been a characteristic of the biblical tradition since the beginning. Yahwism was a sociocultural revolution anchored in religion, a protest against the hieratic states that surrounded the tribes of Yahweh. And early Christians were united in rebellion against the state, primarily the Roman Empire. Of course, the melding of the church (circa A.D. 400) with the Roman Empire changed the course of history, and for nearly a millennium and a half ecclesiastical and political power were effectively one.[4] More recently, the civil rights movement exemplifies the church's influence on the state. The struggle over abortion is another example, but different in that no religiously inspired and politically effective consensus has emerged.

Increasingly, however, the church *appears to have been marginalized*, excluded from influencing public affairs. This split mirrors the fissure between our private and public lives. It is the consequence of a relatively clear sequence of historical events. Both bourgeois society and science itself initially needed religious warrant, but once the modern turn was made, *organized religion* essentially became a dead letter in public affairs. The modern state, modeled on materialistic atomism, has been founded on the metaphysical belief in the absolute nature of individual (atomic) freedom, be it religious, economic, political, or intellectual.

The individual described by utilitarian individualism is, however, a metaphysical abstraction from the social world in which human beings actually exist. We become caricatures of ourselves when we are conceived in terms of a theory that cannot meaningfully describe social relations. Each of us lives primarily within a social context of sustaining relationships, which can be neither empirically described nor theoretically conceptualized as nothing more than the aggregated interests of autonomous individuals. Given the Enlightenment definition of the individual, the "good society" serves private and primarily economic interests. Any vision of the social matrix as constituted by internal relations among people, especially when these are noneconomic, is inconsistent with the prevailing ideology, since relations between two or more people (for example, mother and child) or between generations violates the premise that the individual is (metaphysically) absolute: an atom, in splendid isolation.[5] Americans, whose experiences of relationships are theorized in terms of the language of external relations—that is, market economics—may have difficulty accepting this premise.[6] But John Cobb and Herman Daly argue that a more accurate (empirically and theoretically) way of describing the individual is as a *person-in-community* (1989, 161).

Critics of the prevailing ideology point out that while citizens enjoy freedom from the overt oppression of totalitarian societies (freedom from), they have no freedom to create alternatives to the overriding economic purpose of the corporate state. This lack of positive freedom (freedom to, as distinct from negative freedom, freedom from) is itself a consequence of the narrative that makes the individual a metaphysical abstraction ostensibly enjoying absolute freedom. Society is accordingly conceptualized on the model of utilitarian individualism, where, free from social relations, individuals (either human beings or corporations) pursue their private interests without limit. Society, inevitably, is conceived as nothing more than the aggregate of private interests. Given this Enlightenment narrative, the state can have no other definition than the greatest good for the greatest number of individuals.

The separation of church and state in America exemplifies this liberal-bourgeois theory. Religious believers are guaranteed freedom of worship, that is, the freedom to search for private definitions of the good life. But, on the basis of the liberal tradition itself, the faithful cannot collectively participate in defining social preferences. What this tradition ignores is the fact that on many occasions believers have reached solidarity on the common good (if not so much on specific doctrinal issues). There is nothing to preclude the faithful from coming together in conversation about public issues, such as protection of the biosphere, consistent with the separation of church and state. The word *public* is often misinterpreted as meaning only the enfranchisement of agencies serving the so-called public interest or common good. But public *can also mean voluntary associations of people who seek to define and enfranchise common interests*. Political consensus on an environmental agenda is, I have been arguing, something that might be created out of the diversity of faith, but it does not inhere in faith alone. For it is entirely consistent with the principles of our republican tradition.

The Church in Institutional Context

Living in a political culture founded on the Enlightenment narrative, a story that glorifies the identity and power of the individual, Americans are conditioned to ignore the pervasive effect of institutions on their lives. Any assessment of our culture reveals, however, that we live and work through institutions, and that the shape of things to come will be determined not so much by our deeds as individuals but rather by our actions exerted through institutions. Paul Ekins suggests that at least three kinds of organizational forces are at work in society: (1) "those of the powerful acting in their own self-interest, and possibly against the public good"; (2) "those of government, at many levels, which at least have the chance of being participatory, or representative and accountable"; and (3) "the self-helping, mutually co-operative initiatives of the civil society. People and their relationships will have value only if democratic government and civil society are in the ascendant" (1992, 130). So viewed, the church represents one of the most valuable, if not the most precious, forms of organizational capital in our civil society.

Private and public organizations shape both *social ecology*, our relations with each other, as well as *natural ecology*, that is, our relations to the environment. Consider, for example, either the American system of transportation or agriculture. Individuals lack control over how they move about the earth or secure sustenance. These basic services are controlled by interwoven webs of private (the corporation, such as General Mills) and public (the state, such as the U.S. Department of Agriculture) institutions. As many studies, such as the World Watch Institute's annual *State of the World* reports (Brown 1989, 1990), indicate, the prevailing systems of agriculture and transportation are undercutting ecology. Yet consumers lack another choice. Any satisfactory comprehension of ecocrisis must grasp the organizational context that has created and sustains the problem, yet is necessarily involved in any solution. Environmental dysfunction dramatizes the reality that social ecology and nature's ecology are bound together. Bellah argues that "most of the threats to the planetary ecosystem are the results of habitual human ways of relating to the physical world, ways dictated by institutional arrangements. Inversely, our relations with nature—the way we have used land, materials, and other species—both reveal and shape the institutions through which we deal with each other. But we still have a long way to go in finding a realistic institutional approach to environmental problems" (1991, 15). The vast majority of us, as I attempted to make clear in the Introduction, are "stuck in the system." The issue is not so much personal, that is, good guys with white hats versus bad guys who are exploiting the world, as structural. Given the political will, we can create systems of agriculture and transportation—housing, forestry, energy, and so on—that are economically viable and ecologically sustainable.

Michael Zimmerman argues that a kind of love or relation of respect between people and nature is necessary if we are to heal the wounds. "Such love is necessarily absent in large-scale organizations and bureaucracies" (1992, 175). The large organizations of the corporate state thwart us because they were designed largely to serve narrowly circumscribed economic purposes. Efficiency, as noted earlier, is the watchword of the modern institution, be this business corporation or federal agency. Private corporations serve the bottom-line imperative and the interests of the managers who control them. Public corporations provide the infrastructure—roads and schools, water and electricity—required to make a market society possible. Further, once created and set in motion, institutions are difficult to redirect. Neither private nor public corporations are created, then, to care for creation. By recognizing the almost overwhelming power of these large-scale organizations in our lives, however, the possibility of changing, especially redirecting and more carefully controlling their power, appears.

Generally considered, four institutions shape social and natural ecology: *state, corporation, university, and church.*[7] Between them the state and the corporation are leviathans of such power that they have a disproportionate influence on public life. This reality has been expressed in various ways. Some observers suggest that we live in an era of controlled capitalism or state industrialism. The GNP, the rate of economic growth, and the bottom line are the idols which America worships, and the state and the corporation are the keepers of this faith. The trip to Japan by President Bush and chief executive officers of more than twenty major corporations (1991), including the CEO of Chrysler Motors, Lee Iacocca, might be viewed as a modern-day Crusade—a holy army doing battle with a heathen enemy that threatens our civil religion. In any case, there is no doubt that the "new industrial state," as Kenneth Galbraith (1972) terms it, is constituted by the cooperation of the two most powerful structures of our society: government and the corporation. The new industrial state determines the common fabric of our lives, including transportation, health care, education, agriculture, transportation, defense, and housing.

The State

The state is clearly implicated in any solution to ecocrisis. No analysis can neglect the role that government—local, regional, national—must play. Air pollution, for example, cannot be dealt with apart from public policy initiatives and the organizations designed to implement and enforce them. Neither can

Table 7 Comparison of United States and India (in percentages)

	World Population	Global Energy Consumption	Global CO_2 Emissions	Global GNP
USA	5	25	22	25
India	16	3	3	1

energy or transportation policy. Yet just as clearly, society cannot look to the state alone to solve environmental crisis.

As Vice President Gore tells us, politicians tend "to put a finger in the political winds and proceed cautiously" (1992, 15). Politicians almost inevitably follow rather than lead. Because environmentalism promises to modify the socioeconomic status quo, politicians hang back, afraid to disturb the established order—especially, as noted in chapter 2, powerful private interests. What has been accomplished through environmental legislation and administration in the twentieth century is more a consequence of the dedicated work of environmental groups than the result of any clamor for reform by voters or inspired leadership by politicians (S. Fox 1985; Hays 1987; Cohen 1988; deHaven-Smith 1991). And much of what has passed for conservation is predicated on the notion that nature was a resource that needed to be prudently used so as to extract the last possible measure of economic value from it. Nowhere in the traditional conservation movement does there appear any ecological insight that economic growth for the sake of growth alone can undercut sustainability.

Further, political campaigns hinge on issues of the moment—usually economics. To take a recent example, presidential politics in 1992 turned on economic recovery and deficit reduction, issues the electorate seems predisposed to focus on and politicians and the media are all too ready to serve. Americans seemed to hear only the voices that promised the revitalization of what is already the world's richest economy. The United States of America is not only rich beyond the dreams of most Third World citizens, but is also the world's leading consumer of energy and emitter of greenhouse gases (table 7).[8] As evidenced by the campaign of 1992, Americans seem to want more, and apparently think the only solution to either unemployment or the distribution of wealth is a greater income. The interests of Third World citizens, natural ecosystems, the flora and fauna, and generations of unborn Americans are of little concern: we clamor for a president who will make us better off. Daniel Koshland, writing in *Science*, summarizes this succinctly in relation to the issue of biodiversity, but his point can be generalized to virtually any environmental issue. Politicians will not deny the interests of present voters in economic prosperity in order "to help future constituents who would at best vote for their great-grandchildren" (1991, 717). The half-life of a politician, Koshland continues, is four years, plus or minus two. But issues of ecological consequence are not parceled out in periods of time that correspond with the terms of political office.

Beyond politics we have the actual environmental record of the U.S. government. Much government action masquerades as conservation, though its effect is to push our nation deeper into ecocrisis. The government, through agencies like the Forest Service, Department of Agriculture, and Army Corps of Engineers, too often acts not as the conservator but as the exploiter of nature.[9] Consider also that the federal government is the single largest creator of pollution and habitat degradation in the nation. Environmentalists relate horror stories of the abuse of nature by government officials and agencies in almost every state of the Union.[10] And, as conservatives and liberals alike point out repeatedly, agencies created to regulate private interests for the common good are more often than not co-opted by those private interests.[11] How, the skeptic may legitimately ask, are we to believe that the government can bring order to our culture when its house is in disarray? Even existing environmental legislation—energy conservation, energy research, protection of endangered species, clean air, and so on—is compromised by short-term economic interests. Much of the legislation designed to protect the environment was undone in the past decade. The 1980s have been called "the Reagan antienvironmental revolution" by the environmental historian Samuel P. Hays (1987, 491).[12]

It is difficult not to conclude that all branches of the state are dedicated to the ends of utopian capitalism. Above all else the role of government is to manage the economy: the supply of money, interest rates, the rate of economic growth, the trading of securities, agricultural and foreign trade policies, and so on ad infinitum. These diverse economic functions, necessary to the coordination of a modern economic society, aggregate in the GNP. The GNP is ostensibly an objective (quantitative) measure of social success. The GNP and the rate of economic growth seem to American voters to be a reliable index to the success or failure of our government and society. In truth the GNP is a truncated, obsolete, and increasingly unworkable measure of the economy (Georgescu-

Roegen 1971; Cobb and Daly 1989; Ekins 1992). Ecological economists now offer an array of alternative measures, including the Index of Sustainable Economic Welfare, that provide more accurate measures of cultural welfare. The economic growth imperative, conceived in terms of the GNP, means that issues like acid rain, biodiversity, and climate heating go unmet. It means that the U.S. government will continue to obstruct rather than aid international efforts to reach global agreements on biodiversity and regulation of CO_2 emissions. The issue is not to abandon the market but to empower a more inclusive decision making by society that recognizes the GNP as an index of one kind but not all kinds of wealth.

The dominance of our collective decision making by the GNP merely confirms that utilitarian individualism is the predominating social philosophy. Both the electorate and the government are acting out their scripted roles. Accordingly, institutions that ostensibly exist to protect nature actually serve our generation's selfish interest in economic growth for the sake of economic growth. The first language of American voters, who send presidents and senators and representatives to Washington to set public policy, make the laws, create the bureaucracies, and appoint the officials that enforce the laws, is the gospel of greed. Beset on all sides by the hard realities of a market society, rising taxes, and job insecurity, perhaps it is understandable that Americans should vote their pocketbooks. Economic interests are the conceptual focal point of individuals who perceive themselves as related only through the market. Many observers of the current political scene warn that Americans see themselves almost entirely in terms of individual rights and freedoms. *The public citizen*—the citizen concerned with the well-being of society—*is an endangered species.*

But institutions have no existence apart from language. And institutions necessarily serve the narrative that creates them. No alternative social and natural ecology can come into existence unless it is given voice through a legitimating narrative. Insofar as solutions involve the government this involvement will be through a social consensus developed outside the language of utilitarian individualism. Al Gore insightfully argues that our national "emphasis on the rights of the individual must be accompanied by a deeper understanding of the responsibilities to the community that every individual must accept if the community is to have an organizing principle at all" (1992, 277).

The Corporation

Just as the state, because it is part of the crisis, cannot resolve ecocrisis alone, so, too, the corporation is limited. To expect the corporation to initiate a response to ecocrisis, even under the banner of corporate social responsibility, is naive. Obviously, the corporation is important, shaping the way every need of the American consumer, from transportation to food, is served. Short-term survival for most Americans depends on this system. Yet the negative, long-term implications of environmental crisis imply that producers and consumers are linked in an ecological *danse macabre*. The positive aspect of the problem is that environmental crisis does not indicate any intrinsic difficulty with a market society per se. The environmental disasters that have scarred Russia and Eastern Europe remind us that nature responds to abuse without regard for economic ideology. The problems with corporations have more to do with democracy and the laws governing corporate behavior than with capitalism.

The mass media invariably pick up on corporate horror stories but usually ignore the lessons behind these tales. In Texas, Browning-Ferris, one of America's largest waste-disposal firms, was convicted of spraying deadly chemical wastes on rural roads. No one knows how many people breathed toxic dust, how many animals were poisoned, and what the long-term consequences of these actions will be. What is known is that no Browning-Ferris manager went to jail and that the firm, not its employees, paid the fines. The *Exxon Valdez* has become a national ship of infamy. Again, no corporate official was held either legally or financially accountable. Even more distressing, perhaps, was the ability of the Exxon corporation itself to escape the financial consequences of environmental catastrophe by simply writing off damages against profits.

Of course, the record is not all one-sided. Some corporations do act in environmentally responsible ways. Public opinion sometimes influences corporate behavior, as in the case of McDonald's moving from styrofoam to recyclable paper containers.[13] Corporations sometimes move ahead of public opinion and take the lead, as in the case of Mary Kay Cosmetics. Mary Kay initiated voluntary corporate recycling and donated the proceeds to environmental organizations. And corporations can be good public citizens, attempting to deal with the issues of concern sometimes at the expense of the bottom line. Chevron, for example, one of the largest energy companies, continued to include an expensive fuel additive in its product, even after government regulators barred advertisements of the additive, because in-house research had shown that it improved fuel economy. Electric utilities in California are setting a national precedent by emphasizing energy efficiency over construction of additional power generation facilities. And seventy-five industrial plants in Texas have publicly pledged to reduce toxic emissions (at their own expense) by 60 percent by the year 2000.

Still, these actions are basically stop-gap. In the present ethical and legal environment, the bottom line tends to short-circuit any potential the corporation has to redirect American society toward sustainability. Corporations are followers, not leaders, mired in the first language of utilitarian individualism. Adam Smith and Thomas Jefferson did not envision the giants that now rule the business world. Their vision was of small merchants, farmers, and entrepreneurs—proprietorships, not corporations, owned and operated by individuals living in place, anchored in a biblical tradition. Neither could Smith and Jefferson envision the competitive environment in which the modern corporation exists. Some environmentalists naively charge that there is a ruling elite of corporate insiders who control the government and forestall environmental reform. The empirical reality is that the competitive environment itself makes it difficult for one corporation to take environmentally responsive action. The manufacturer that voluntarily installs pollution abatement equipment will soon be out of business, since profit-maximizing rivals will enjoy a competitive advantage. Environmentalists argue that voluntary, industry-wide agreements could establish pollution standards and avoid bureaucratic costs, but *competition also undercuts voluntary cooperation across firms*. Corporate secrets are usually guarded jealously. Automotive emissions control standards and technologies, for example, were slow to develop in the 1970s and 1980s because of competition. Discussions between automotive and energy corporations were thwarted because "emissions control design has become an inherent part of overall vehicle powertrain design, and therefore a closely guarded secret in the highly competitive auto business" (Spitler 1992, 112–13).

Corporations, however, are not part of the basic furniture of the world; more than anything else they reflect the Enlightenment narrative. It follows that corporations can be changed, that is, placed within the context of a new legitimating narrative. There is good reason to reconsider Milton Friedman's (1962) arguments that political and economic freedoms go hand in hand, and Galbraith's arguments (1952, 1972) that the merger of the corporation with the state in the new industrial state threatens basic democratic freedoms. The stubborn reality remains that we do not have time to reinvent the wheel. Business corporations will be involved in the quest for solutions to ecocrisis: *the key is to find ways to redirect corporations in environmentally responsible directions*. This will come to pass only when an enlightened electorate has reached solidarity on caring for creation. Such a consensus might (1) mandate environmental audits of corporate activities, (2) place public directors on corporate boards, (3) establish industry-wide standards for pollution control, (4) make corporate managers criminally and financially liable for corporate malfeasance that adversely affects the environment, (5) preclude tax write-offs for the costs of environmental clean-up and fines, and (6) protect whistle blowers.

The University

A third major player in the institutional mix that characterizes American society is the university.[14] Institutions of higher learning and affiliated research institutes trumpet resounding claims as to how the "knowledge revolution" or the latest "scientific research program" will transform society. Some members of the university elite have announced modest plans to "manage planet Earth." Yet faith in our universities to promote structural change vis-à-vis ecocrisis is difficult to maintain. A few, such as Galbraith, hold out hope, arguing that since the new industrial state requires professionals, and since the university trains these individuals, the system can be redirected from within. Galbraithian optimism is difficult to sustain for anyone familiar with the inner workings of the modern university, private or public. The research university less and less leads society in new directions and instead more and more reflects society's established characteristics.

Increasingly, the university is run as a corporation: with managers who, as they allocate academic funds, respond to the needs of the corporate job market; with professorial entrepreneurs who cater to the market for fundable research, driven historically by the U.S. Department of Defense and corporate research and development; and with oversight groups concerned largely with economic efficiency. In an era of declining funding, universities are continually pressed to do more with smaller budgets. Administrators, from the level of chancellor to departmental chair, are driven to make decisions on the basis of financial constraints rather than educational objectives. Classroom enrollments burgeon in size, justified by claims that larger classes boost productivity. If one professor teaches twenty students then, by this logic, that same professor is twice as productive with forty students in the classroom—or eighty. Even worse, teaching assistants, long an exploited source of inexpensive teaching help, have assumed full control of even upper-level courses. Prestigious institutions of higher learning proclaim vaunted core-curriculum reforms that on closer analysis turn out to be cosmetic changes designed to win outside funding.

Some look to one segment of the university for a vision of the future—to environmental studies, such as restoration ecology, climatology, conservation biology, and environmental philosophy. No doubt such expertise, research, and academic study are useful and necessary. But the possibility of achieving a sustainable condition of existence from the initiatives of environmental studies are few to none. For one reason, as Neil Evernden explains, ecology is more part of the discourse of utilitarian individualism than of any effort to seek cultural transformation. He argues (1992a) that ecology as traditionally conceived is implicitly technological, designed more to increase the human drive to dominate the earth than to help humans live in harmony on it. An ecological voice fostering the second end might someday be heard, but it is at present quiet.

Further, the university's effectiveness as a mediating institution has been undercut by its organization into disciplines that isolate themselves methodologically—a "sin" that has been called "disciplinolatry." Academics have become so specialized that members of the same discipline sometimes find themselves unable to follow their colleagues's conversations. Mathematics, for example, is so fragmented into specialties—more than two hundred—that a mathematician trained in more than two or three is considered a polymath. The situation is similar in the social sciences, where jargon is so byzantine that one may wonder if even insiders comprehend one another. The consequence is that the public sphere has been impoverished, derived of the sustaining flow of intellectual reflection and analysis that nurtures it.

More generally, universities have become little more than *associations of individuals*—students, faculty, administrators—in pursuit of their own selfish ends. Few in the intellectual world are capable of a synoptic view. The professoriat, though it gives lip service to teaching, generally knows where its bread is buttered: specialized research, publication, and grants. Those faculty who wish to invest in quality teaching are increasingly dispirited, whipsawed by declining budgets and more entrepreneurial colleagues. Students, too, are under the spell of utilitarian individualism, seeking not the traditional end of knowledge as an end in itself but marching to the dictates of economic success. A university degree is not a guarantee of financial success, but the road to wealth usually requires it.

The Church

Assuming, then, that the state, the corporation, and the university are incapable of leading our culture toward sustainability, however much they might be involved in that transformation, we are left with a single alternative: the church. Setting the church within the larger institutional framework clarifies the role of religion in a time of ecological crisis. Not only can religion help us begin to appreciate the earth as something more than a resource to fuel manufacture and as an environmental sink for pollution, but the church itself—regardless of denomination—is, or at least can be, the organizational vehicle of cultural innovation. At this juncture in history, religious discourse offers the most accessible alternative to the language of utilitarian individualism, which holds the state, the corporation, and the university in its sway.[15] True citizenship ultimately involves the binding of individuals in collective purpose, something beyond mere economic greed in which I get mine and you get yours and we are all better off for it. *The church, then, is a necessary part of the politics of community, for it can sustain a dialogue that can take us beyond ecocrisis.* The church is a mediating institution that can create a space for collaborative discourse on an environmental agenda—a place where the rhetorical tradition and citizen democracy can be revitalized.

Admittedly, the church has a legion of institutional liabilities of its own. As Cobb points out, many individuals are "tired and confused by the endless demands for change and want the church to be an island of confident changelessness in the sea of secular confusion. For them the old-time religion is the answer" (1990, 264). Like many others, Cobb recognizes that such a position is self-defeating and may ultimately lead to a cultural collapse. He thinks that the multiplication of "theologies of"—such as theologies of women's experience, of liberation, of work, and so on—is both a danger and a promise. The promise is inherent in the creative agency of such theologies, as they raise voices of change. The danger is that the church "will incorporate only what can be assimilated into the mainstream of a relatively unchanged tradition" (264). But the church remains the one institution where habits of the heart, the language of the community, yet exist.

The Church and Habits of the Heart

Granted the power of organizations, we readily see that the corporation and the state move modern society on a relentless trajectory of materialism, of economic growth for its own sake. Religious discourse has been marginalized in public affairs by the gospel of greed. How, the skeptic asks, can the church make a difference? The church is involved, I believe, if for no other reason than the reality that our ethical and political lives are collapsing as the earth is destroyed. As we disrupt the web of life in the unbridled pursuit of a narrowly defined and scientifically tenuous conception of economic success, the fabric of our culture is unraveling. There seems to be no hope as our world plunges toward global ecocatastrophe. The individual, embedded in the institutionalized order of life, can do little to change the course of events. And collectively we are driving nature toward a point where civilization will no longer be sustainable. This is not, I have argued, truly surprising. The modern person has lost sight of the sacredness of creation. Is there any wonder that in becoming Homo oeconomicus we are endangering the Creation? But the real question is: Can the profane person of the modern world again learn to care for creation?

Perhaps. As chapter 4 implies, there is *reason for hope*, a promise of renewal across the spectrum of contemporaneous religious belief. And the importance of finding hope in organized religion should not be underemphasized, since it "rests on something other than its own usefulness" in a time of ecological crisis. Hope, John Cobb argues, is not so much a function of its utility as "a function of what we believe, and in this cosmic and global crisis, it is most clearly a function of what we believe *ultimately and comprehensively*" (1992, 124). It is easy for the overwhelming power of the established order of things to vanquish us psychologically. But for believers, Cobb explains, the experience of the Spirit in themselves calls "forth the realistic hope apart from which there is no hope" (125). The religiously faithful can find, especially in the sustaining community of faith—the local congregation itself—the strength of conviction and the power of hope. With the telling of the sacred story comes the renewal of Spirit. In spite of the diversity of religious belief, there is a common ground for caring for creation that can make organized religion a political turning point. Renewal begins within *the local church*—the immediate faith community to which individuals are committed. The local church is the key place to shaping opinion because (1) the church has had and retains importance as a *mediating institution* between individualism and large institutions and because (2) every church has within its own narrative tradition the power to challenge the language of selfishness and the gospel of greed.

So construed, the church is a fundamental political institution whose function is not to set policy but to support the principles that make democracy possible and to encourage concern for others, including future generations of human beings and the rest of the Creation. Religion does work to shape political attitudes and behavior. President George Bush, for example, invoked religion to justify the righteousness of Operation Desert Storm. Yet it is at this cultural level, as we have seen, that liberals become suspicious and controversy arises over the intrusion of religion into public life. Liberals prefer to think that elected officials and the government more generally stand for the morally right thing, since they enjoy majority support. This is a fallacy, since majority support does not confer morality on governmental power. It is a truism of democratic life that the majority does not necessarily know what the best policy or candidate is. The sorry history of National Socialism and Adolf Hitler, who was installed by a majority of the German electorate acting under the Weimar Constitution, is proof enough. The belief that the majority opinion is intrinsically moral overlooks, according to Abraham Kaplan, the political reality "that consent can be cajoled as well as coerced; virtue is lost to seduction more often than to rape" (1963, 76). Kaplan's remark implies that utilitarian individualism seduces Americans into supporting the economic status quo, the industrial growth paradigm. The irony is that an ecologically sound economy would be good for everyone. Caring for creation entails not economic suicide but rather the embracing of economic principles that go beyond merely monetary measures of welfare.[16]

Americans appear to be increasingly skeptical of the results of the electoral process (as distinct from the idea of democracy), politicians, and the policy-making process. The National Commission on the Environment reports that "the percentage of Americans who said they trusted the government 'to do what is right' always or most of the time declined from 76 percent in 1964 . . . to 28 percent in 1990. During the same period, the percentage of Americans who thought that the government is 'pretty much run by a few big interests looking out for themselves' went from 31 . . . to 75 percent" (1993, 52). Americans are suspicious in part because their votes do not appear to make a difference. Whatever their promises, politicians seldom act in ways that change cultural outcomes: the status quo rules. When politicians drape themselves in flags, stand on podiums framed by the Grand Tetons, and declare themselves "environmental presidents" while their advisers declare that "methane is not a greenhouse gas," something is fundamentally wrong. Skepticism about the political process is not misplaced. The distrust also dramatizes the potential of the church to facilitate the democratic process leading to solidarity on an environmental agenda and, subsequently, to adaptive political responses. More precisely, in serving its role as a mediating institution, the church positions itself at the juncture of politics and ethics. Which is to say that the biblical tradition reinforces the republican tradition—the heritage that seeks the public good and ultimately, through communal discourse, attempts to develop a consensus on the public good.

For most Americans *the local church* is far and away the most likely forum for discussion of moral issues that overlap with politics. The local church is ideally suited to discourse where ecological crisis runs up against the gospel of greed (and where ecology as objective science is reluctant, either incapable or unwilling, to assume a normative stance; see chapter 2). The church is a community of memory, tracing its roots back to a covenant relation with God and the celebration of that relationship on the sabbath. The worship retells (better, recreates) the story of the relation between the religious community and God, and the liturgy, ideally, provides a legitimating narrative for the whole of life. Americans have reasons to care for creation. Insofar as these traditions remain viable, they are not confined to a day of worship but shape the believer's character and behavior, spilling out of the church into everyday life.

More fundamentally, the local church starts the body politic moving toward a working version of *the public church*. As Martin Marty (1981) argues, the public church refers not to any actual institution or denomination but to a "communion of communions" that remains consistent with the diversity of faith traditions. The communion of communions flourishes wherever issues of the public good exist—ecocrisis being an obvious example. That the public church exists is beyond doubt. Organizations such as North American Council for Religious Education (NACRE), the North American Conference on Christianity and Ecology (NACCE), and the World Council of Churches (WCC) testify to its reality.

Consider a few examples. NACRE organized a national conference on Caring for Creation, celebrated during May 1990 in Washington, D.C., that was attended by more than 2,500 delegates, including large numbers of clergy as well as representatives from the scientific community. Portions of the proceedings were televised nationally. The WCC has also been very active on issues involving global ecology, attempting to work out ecumenical policy statements for the Protestant community. In February 1991 the WCC held an international conference in Canberra, Australia, with the theme "Come Holy Spirit, Renew Creation." Particularly promising in regard to global ecology is the report to the WCC entitled "Liberating Life: A Report to the World Council of Churches."[17] This report calls for developing a theology for the liberation of life. "Informed by the biblical witness, the insights of science, and our experience of the interdependence of life, this theology needs to address the brokenness of our world and its intricate web of life with a new statement of the healing words of Christian faith" (Birch 1990, 276). The WCC was also present at the Rio Summit (or, more accurately, at the alternative summit held outside the "governmentally sanctioned" activities), where it advocated the theme of "Justice, Peace, and Integrity of Creation."[18]

Beyond the WCC, an organization supported by most Protestant denominations, are denominational statements on ecology and the Creation. As noted, American Catholicism (see chapter 4) has spoken authoritatively on the issue of ecocrisis; so has American Protestantism. Most American Christians, both Catholics and Protestants, are in churches that have adopted ecological position statements and have active ecological ministries. According to Massey (1991), among the thirty largest Protestant denominations, the only denominations that have no such policy are the National Baptist Convention (fourth largest) and the Church of God in Christ (seventh largest). As noted, the only denomination that has formally stated its opposition to ecology as part of the church's mission is the Church of Jesus Christ of Latter-day Saints. Collectively, however, these position statements indicate a commitment to the public church—to a communion of communions that cares for creation.

In addition to national denominations and international organizations, many regional and local organizations carry on the mission of the public church, including Jewish, Protestant, and Catholic groups. *Shomre Adamah* (Keepers of the earth), for example, provides ecotheological information and books for people of the Jewish faith. Headed by Ellen Bernstein, the organization distributes traditional (selections from the Pentateuch) and nontraditional educational materials. *The Grassroots Coalition for Environmental and Economic Justice*, organized by a former Jesuit priest in 1988, operates in Maryland. Its aim is "to encourage and assist church members to work effectively for environmental justice for our Earth-Community." Members of the grass-roots coalition work with local churches (on invitation) to initiate a process of caring for creation within the congregation.

Interestingly, although the ecologically oriented activity of the public church is increasingly evident, the tradition of Judeo-Christians forming organizations that attend to the Creation dates back to the 1930s. Even as early as the 1940s, according to Rod Nash, some churches were exploring their obligations to the natural world. In the wake of the Dust Bowl, the National Catholic Rural Life Commission brought religion to bear on the issues of soil husbandry. "The observance of Rogation Days, which dates to the Middle Ages and acknowledges human dependence on planting and harvesting, provided a conceptual basis for modern dedications such as Rural Life Sundays and Soil Stewardship Sundays" (1989, 98). And in the 1950s, Nash continues, the National Council of Churches "launched a program called 'A Christian Ministry in the National Parks,' but its emphasis was largely on human appreciation of the beauty in God's world" (98). During the 1960s and 1970s theologians formed a variety of organizations and study groups to ground an environmental ethic in religion. The earlier work of theologians like Joseph Sittler, Charles Hartshorne, and Daniel Day was very influential during this period. During 1963–64 the National Council of Churches formed the Faith-Man-Nature Group, which was dedicated to articulating a Christian environmental ethic, and it held annual meetings from 1965 until 1974 (until financial exigency ended its all-too-brief tenure). The recent activities of both national and local religious organizations, then, are not unprecedented, but what had been largely episodic and fitful during from the 1930s through the 1960s and even 1970s is now a steady and growing stream of activity. Put in slightly different terms, the work of the believers who cared for creation during the 1960s and 1970s is bearing fruit. What was for too long nothing more than a few isolated efforts shows signs of becoming a powerful social movement.

The Local Church

It is no secret today that the corporation, government, and university have lost touch with the vital interests of the mythical yet nonetheless real average person. The church has not. The United States remains a nation of religiously affiliated people, most of whom are active in their churches. Statistical surveys have discrepant results, but Theodore Caplow's *All Faithful People* is a useful study; according to Caplow, 95 percent of us believe in God or Deity, 60 percent are church members, and 40 percent attend a service at least once a week. The church ties Americans to their Judeo-Christian past; it roots them in the biblical tradition that helped to initiate and may help to sustain Western culture. From the beginning Judeo-Christianity was a covenantal faith, and many Americans believe that their communities of faith still stand in a covenant relationship with God. Judeo-Christian worship services celebrate that tradition, renewing and celebrating the relation of the community and the individual with God, tying the individual to the sacred canopy. Through storytelling church members become part of a *community of memory*—reinforcing their sense of history and thereby becoming more effective members of society.

The church, of course, has quite different meanings depending where individuals stand on the spectrum of belief (see chapter 4, figure 2). For some, the church is understood as exemplifying the living Christ. For others, such as nature religionists like John Muir, the church is the earth community. But for all believers, the church functions as a corporate entity, offering individuals a meaningful social context that endures beyond their lives. For individuals the church's durability is a stable center in an otherwise transitory life; its community helps them escape solitude. The church is a living web of belief—narrative, symbol, liturgy, prayer—that sustains people, integrating them with sacred story. By participating in their community of faith, believers gain a sense of legitimacy, of authenticity that goes beyond the merely contingent identity provided by modern culture.

Caring for creation is clearly an appropriate theme for worship in local churches, regardless of denomination. Creation stories are central to this effort. Worship reminds people of their relationship to God and to the sacred canopy. By placing human life in an order of meaning beyond utilitarian individualism, worship builds character and virtue that can work to transform economic and political life. Prayer and liturgy, too, create a sense of belonging, of communion with a living God. Worship, liturgy, and prayer vary widely across the spectrum of faith, reflecting a particular creation story and ecotheology. Conservatives (see chapter 4, figure 2) might build a service around Deuteronomy 8.7–11 to emphasize the goodness and sacredness of God's creation. The passage underscores the idea that the land nurtures humans. And it leads to the question of obligations to the land, to the Creation itself. Goddess feminists, at the other end of the spectrum, might choose a passage from Susan Griffin's works or a poem by Emily Dickinson to accomplish the same end. These writings proclaim a message similar to that in Deuteronomy, powerfully and evocatively describing a sacred earth that sustains human beings.[19]

The local church can also care for creation in ways that go beyond worship and liturgy to include religious education.[20] Basic elements of programs for elementary students, junior and senior school students, and adults are suggested below (figures 3, 4, and 5). As in chapter 4, my discussion merely introduces the subject of religious education that builds an environmental ethic, a caring concern for the Creation across the congregation, from young children through senior adults. Educational programs necessarily have a focus consistent with the faith tradition of the local church. But the presentation of scientific information about various aspects of the environmental crisis is also important. Further, a religious education program should attempt to involve every member of the congregation. The congregation should try to be part of the solution for environmental crisis.

We should begin, appropriately, with the children, since the gospel of greed is stealing their blessing (figure 3). Youngsters can make important contributions to the congregation; they can, for example, help their families plan ways of living that are less harmful to the environment. Children can be quite candid, and they often cut to the heart of the matter, avoiding adult subterfuges and stereotypic ways of thinking (see Isa. 11.6–9). Youngsters who

Figure 3 Church-related Educational Activities for Youth

Bible stories

Noah's ark and the lesson of stewardship. Parables concerning Jesus's love and care for creatures.

Show and tell

Children might bring pets (cats, turtles) to class, discussing the care and health of the animal and human responsibilities to wild as well as domesticated animals more generally.

Nature walks

Introduce the local flora and fauna; the beauty of fall colors; the mystery of an ant hill; the wonder of a bird's nest; the interrelations between sun and shade and plant growth; the effects of the wind. In urban areas, trips to a local park may be appropriate.

Home activities

Encourage children to think of ways that they can help nature in their home and daily living—by turning off lights, recycling cans and paper, conserving water, and so on.

---

participate in church-related educational activities gain valuable learning experience. Children learn that they can make a difference through their own actions. They will also acquire a better perspective on environmentalism—learning that even little things, like recycling in one household, can mean a lot in the aggregate. Most important, they will learn that their faith is a living faith, something that makes a difference each and every day, not just on days of worship.

The education of teenagers presents a wider array of possibilities and challenges (figure 4). In part this is because adolescents are more deeply socialized in the ways of culture. The language of materialism and consumerism is beginning to influence their behaviors and aspirations. Characteristically, they want what their parents and peers have, only more of it. The problematic and unresolved issues of American culture flood over them—drugs, alcohol-

Figure 4 Church-related Educational Activities for Teenagers

Bible reading

Instructors should select passages for student consideration and interpretation or application as appropriate.

Fact finding

Community-related issues are effective in attracting the interest of young adults. Holes in the ozone seem abstract, but landfill issues, energy conservation, or ecological restoration projects are closer to home.

Service projects

Planting trees, holding environmental poster fairs, conducting recycling drives, and raising funds for environmental causes are valuable exercises. Everyone learns to do something.

Discussions

Conduct sessions in which teenagers can discuss with candor the mores of their group. Draw out the environmental consequences of the American way of life; explore what your tradition of faith provides as an alternative.

Home activities

Encourage young adults to be imaginative in designing ways to make households greener (for example, meal planning and preparation, heating and cooling of the home, transportation).

---

ism, sex, materialism. The peer group begins to assume even greater power of influence. With some exceptions, adolescents will find a direction for their life that mirrors that of their peer group. The peer group itself is shaped by society, especially through the media and popular culture. Teenagers are bombarded with and often overwhelmed by images of the good life—sex, money, fame, automobiles—and important people—athletic superstars, rock and movie stars, powerful business leaders, and politicians. Through religious education, the church can provide a psychologically powerful, alternative socialization that offers other visions of the good life and carries the message of caring for creation. Youths can become involved in projects that make a difference locally, empowering themselves in doing so and rejecting the sterility of "superstars" as models.

However important the education of children and teenagers may be, it is adults who have the ability to affect immediate changes in economic and political behavior. Children are tomorrow's voters, and they don't control the family purse-strings. And parents do more to socialize their children than either the church or the school. Thus, religious education for adults is essential to any pragmatic plan to care for creation. Ten suggested activities for adult education (figure 5) run the gamut from Bible reading to ecological inquiry and action plans to ecotheology. Granted, the power of religion to hedge in narrowly economic interests has waned steadily since the nineteenth century. Privatization, argues Tipton, places "religion, together with the family, in a compartmentalized sphere that provide[s] loving support but . . . [can] no longer challenge the dominance of utilitarian values in the society at large" (Bellah et al. 1985, 224). But this pessimism threatens to become a self-fulfilling prophecy. The public church is within reach. And the prophetic tradition marks the path toward sustainability and learning to care for creation (see chapter 6). Whatever else it may do, religious education can invigorate a congregation through conversation that makes faith relevant to the conduct of daily life.

The Public Church in Theory

I am now in a position to extend the argument concerning religion and environmentalism to the claim that local congregations collectively influence the body politic through the public church. Obviously, by hearing, narrating, and discussing creation stories within and among congregations, the public church at least starts to come into existence. Yet the existence of the local church complicates claims about the public church and its function. How can theologically diverse, highly individualistic, geographically separated bodies of worship achieve a consensus on the environment—on green politics and policy? The weakness of the social gospel movement, the heavy retrenchment of parachurch organizations during the 1980s, and the general malaise of the ecumenical movement suggests that coordinated action at a denominational let alone ecumenical level will be difficult to achieve. Further, the diversity of traditions of faith seems to undercut any claim that solidarity on the environment can arise through the public church. As James Nash (1991) argues,

Figure 5 Church-related Educational Activities for Adults

Bible reading and interpretation

Assign members of your study-group to find passages relevant to caring for creation. Use these for meditation, prayer, and group discussion. What specifically does the Bible tell you about your responsibilities in a time of ecocrisis?

Fact finding

Assign members of your study-group to develop a bibliography of relevant reading, and to gather information on issues of the global ecocrisis, such as biodiversity, climate heating, population growth, and old-growth forests.

Church issues

Conduct an ecological inventory of your place of worship. Can energy be saved by investing in efficiency? Can materials used in church activities be recycled? Is the church part of the problem or part of the solution?

Household changes

Inventory the possibilities for greener ways of living in your household; involve children in this activity. Focus on utilities, garbage, food selection and preparation, and transportation. Share your experiences, successes, and failures with the group.

Expert testimony

Invite speakers from universities, governmental agencies, and environmental groups to interact with your study group.

Congregational action

Evaluate your denominational statement on ecology. Is it a good policy? Can it be improved? How can it be implemented? Assign one or more group members to develop bibliographies of relevant readings. Send delegates to the next regional or national meeting of the denomination. Get green issues on the agenda.

Community planning

Evaluate the community master plan. What are city planners doing to protect the

Figure 5 *Continued*

environment and conserve the quality of life? Explore local building codes, transportation master plans, utility rates for heavy users, land use policies, recycling policies, and so on. Identify areas where the church can facilitate community action.

Advocacy and political action

Make it clear to your children that you are concerned with caring for creation. Explore the curriculum at your local school; make sure students are receiving environmentally relevant information. Form advocacy groups prepared to share the work on vital local issues and elections. Evaluate all candidates for local, state, and national office. When green advocates constitute a democratic majority then *all candidates* will be green. Be skeptical of candidates with empty environmental rhetoric. Ask for specifics from politicians.

Ecotheology

Survey ecotheologies from across the spectrum of faith. Which appeal to you? Which do not? Can attractive elements from other ecotheologies be incorporated into your tradition?

diversity on the specifics of environmental issues is to be expected, population policy being a prime case in point.

One answer is to think of the public church as achieving solidarity within diversity rather than as being monolithic. Marty (1981), as noted above, defines the public church as "communion of communions," as a vehicle for the discussion and articulation of new conceptions of the public good that go beyond religious individualism and tribalism. The public church is not a body or organization where different factions concur on policy recommendations or even are presumed to agree on all issues. Some observers suggest the image of Jacob's ladder to visualize the relation of local churches and denominations to the public church. The implication is that solidarity on an environmental agenda does not require theological orthodoxy. Tipton argues that "the public church is not triumphalist—indeed it emerges in a situation where Christians feel less in control of their culture than ever before—but it wishes to respond to the new situation with public responsibility rather than with individual or group withdrawal" (Bellah et al. 1985, 239). *But if local churches learn to care for creation, a political consensus can be realized that is entirely consistent with religious diversity.* The public church enables solidarity on an environmental agenda by coalescing on *social preferences* that are distinct from *consumer preferences* expressed through the market. Such a consensus neither logically requires one ecotheological rationale nor politically entails a state religion. As I illustrated in chapter 4, the same point can be reached by many routes. The public church includes all people of faith across the ecotheological spectrum.

Tipton argues that "if there is to be an effective public church in the United States today, bringing the concerns of biblical religion into the common discussion about the nature and future of our society, it will probably have to be one in which the dimensions of church, sect, and mysticism all play a significant part, the strengths of each offsetting the deficiencies of the others" (Bellah et al. 1985, 246). Catholicism alone sweeps the spectrum from church to mysticism—from the pronouncements on ecocrisis of Pope John Paul II to the creation spirituality of Matthew Fox. Marty maintains that even though sects making "tribalist claims," "catering to private interest," or claiming "special interiority" have an advantage over the public church in terms of maintaining themselves as institutions, the public church can not only survive but grow stronger (1981, 167). One reason, as he puts it, is that "the choices are stark: a bleakly secular landscape or a belligerently religious one or, more likely, a continued mix of both. Over against them, however, we can compile some assets. Countless congregations in the public church *do* work effectively" (169).

In the context of the public church, it is important to distinguish between thinking theologically about nature and discourse within a church. The lack of any one ecotheology to which every individual, sect, or denomination subscribes is not a barrier to caring for creation. The lack of a master narrative does not preclude solidarity on an environmental agenda. Cobb (1990) touches on this issue in his important essay "The Role of Theology of Nature in the Church." He argues that in a time of ecocrisis what is most vital is not the articulation of ever more clever and lucid ecotheologies but rather the discussion, elaboration, and celebration of existing creation stories by existing church communities. Many diverse cosmic creation stories already exist. These stories effectively constitute a second language, a moral discourse that checks, even redirects the first language of the state, the corporation, and the university.

The local church is a point of initiation for shaping voter preferences that lead to public policies addressing ecocrisis. The church is where individuals, caught up in the flow of their lives, in their joys and sorrows, can begin to grapple with the complexities of social existence. The church, in other words, is the natural home for dialogue that centers on creation, our place in the Creation, and our obligations to creation. The church is a place for conversation, where the faithful tell and retell stories that are emotionally evocative, psychologically persuasive, and ethically charged. Religious discourse can also lead individuals beyond the denomination or sect or mysticism toward conceptions of the public good. It is the church, in its function as a public church, that will help make environmentalism a reality. We should try to think of the public church as a coalescence among all traditions of faith on a common agenda to care for creation. Although such a convergence of public opinion has not yet produced even a politically effective minority, as virtually all commentators on environmentalism recognize, the potential for solidarity is there. The continuum of creation stories implies that a *culture of coherence* is within reach, a culture that might succeed our present *culture of separation.*[21]

Conclusion

A little more than a century and a half ago, Alexis de Tocqueville came to America. On his return to France he wrote *Democracy in America,* a book that has assumed the proportions of a classic. In 1985, Bellah and his coauthors, in *Habits of the Heart,* brought Tocqueville into relation with the present. Tocqueville's crucial insight, apropos of my argument, is that the genius as well as the Achilles' heel of democratic life is freedom itself. Tocqueville feared that guarantees of individual liberty might erode the common purposes and commitments necessary to sustain democratic society. The saving grace in such a society, he argued, was the existence of mediating institutions, such as the church, voluntary associations that bring individuals together in the quest to define public values and thereby link them to the national government.

Assuming the existing framework of liberal democratic society *and* capitalism, since any wholesale revolution in the basic institutional structure is unlikely, the role of the church is clear. Religious freedom, a vital component of our liberty, can simultaneously check unfettered economic greed, without eliminating the market, and shape public values, their articulation as well as their ultimate realization through political process. *Sustainability and capitalism are not incompatible.* Sustainability does entail public policy initiatives, but no credible argument exists to the effect that sustainability entails state economic control. The record of socialist societies on environmental policy is abysmal.

Many responsible commentators believe that statism is in many ways responsible for ecological crisis—that government is more a part of the problem than the solution for ecocrisis. But sustainability (that is, the resolution of ecocrisis) cannot be achieved in a society in which private values determine all economic behavior, either productive or consumptive.

The church is the most important mediating institution in American society; but recognizing the potential of the church to mediate between private and public values is one thing. Making that potential a reality is another, because the church is embedded in an individualistic culture. The diversity of faith in America is no barrier to solidarity: there is no evidence that theological consensus is necessary to produce an effective political consensus. The diversity of creation stories in fact confirms that there remains, among all traditions of faith, legitimating narratives beyond utilitarian individualism. Still, the temptation to withdraw into a closed circle of private interest and devotion, or tribalism, as Marty terms it, is a constant threat to the existence of the public church.

MacIntyre, for example, appears to retreat into tribalism, writing that "what matters at this stage is the construction of local forms of community within which civility and the intellectual and moral life can be sustained through the new dark ages which are already upon us. . . . This time however [unlike the collapse of the Roman Empire at the beginning of the Dark Ages] the barbarians are not waiting beyond the frontiers; they have already been governing us for quite some time" (1984, 263). MacIntyre correctly sees that the managers of the new industrial or corporate state overdetermine us all. In truth, almost everyone recognizes this, as reflected in the declining trust Americans have in government to do the right thing. But if MacIntyre is correct, then the local church will become merely a haven, a sheltered harbor, a green spot in a cultural desert that undercuts its potential to participate in the public church. The larger issue is whether religion itself is succumbing to the Enlightenment narrative and church membership is becoming another means to the end of self-fulfillment.

One answer is that religious commitment and public involvement are intrinsically incompatible—virtual contradictions in terms. The other is that religious devotion and civic concern are not intrinsically at odds. Granted, the church is a source of identity for individuals, who find their places within communities of memory. But this world opens onto a public world. By participating in the ongoing conversation that is the biblical tradition, individuals sustain that tradition and find meaning beyond their own egos. *Nothing seems more antithetical in principle to the historical inspiration of the biblical tradition than insularity, exclusivity, and privacy.* The apparent loss (at least to most churched Americans) of a theological rationale for any church beyond the local church is countered at the theological level by a diversity of ecotheologies that care for creation.[22]

Ultimately, at least in relation to ecocrisis, all religions can find their place within the public church. Insofar as any claim that Judeo-Christianity remains capable of offering moral guidance and direction in a time of ecocrisis is defensible, then the public church must exist. Diversity is a fact of life, but the diversity of traditions of faith need not preclude a "public theology" that cares for creation. The public church lives in and through diversity. Each tradition articulates its own creation story. But all find solidarity in a common core concern of caring for creation. The public church can ultimately be seen as a mediating institution that helps American culture articulate social preferences that cannot be expressed through the market.

A final analogy is perhaps useful. In the 1960s the United States was torn from coast to coast by race riots. No one explanation suffices, but lack of educational and economic opportunity and the resulting poverty and sense of utter despair were important factors. In 1992 the United States was again torn by riots, ignited by the now infamous trial of Los Angeles police officers accused of beating Rodney King. Fact finding revealed that a lack of educational and economic opportunity and the resulting poverty and sense of utter despair were again contributing causes. The one difference was that these conditions were worse in the 1990s than they had been in the 1960s.

In the Introduction and in chapter 1, I argued that, despite two decades of effort and large expenditures of capital, most of the salient indexes of the degradation of nature are now worse than in the 1970s. Our natural ecology is in tatters. By all appearances a similar paradox occurs in social ecology as well. Politicians can only lead a democratic society where the citizens wish to be led. The political establishment serves the status quo. Surely no informed citizen can doubt the truth of this claim. If new directions are in order, then those preferences must first be articulated by the citizens. Above all else, America's social and natural ecology continues to worsen because of the poverty of its public philosophy: utilitarian individualism. We live in a Republic of Special Interests: private interest (which is, more than anything, a euphemism for greed) overdetermines us all. The public church is where the political infrastructure, the basic democratic armature that drives our society, can reestablish itself. I have limited my argument to the environmental agenda; but surely the social agenda is included as well.

6

Redescribing Religious Narrative: The Significance of Sacred Story

It is hard to picture getting a spiritual revolution going without reverting to the story, the history, of the community. . . . Conversely: do not worry, over the long pull, about people who try to change the world without "story." They may project utopia, but getting from here to there demands attention to event and events. People are relatively powerless if they lack identity, plan, or plot as grounded in history.—Martin Marty, *The Westminster Tanner-McMurrin Lectures, I*

Ecocrisis and our industrialized, high-consumption, profit-maximizing economy are clearly not unrelated. But economic affairs affect more than the earth. Henry David Thoreau, writing in the nineteenth century, offered an insightful critique of the prevailing economic paradigm. *Walden* advises that most people are living lives of quiet desperation, "so occupied with the factitious cares and superfluously coarse labors of life that its finer fruits cannot be plucked by them" ([1854] 1962, 109). Thoreau's biting essay "Life without Principle" concludes that nothing is more opposed to life than the incessant pursuit of wealth. Thoreau is restating cultural constants. The Great Code, which surely influenced him, indicts the pursuit of wealth as an end, not only for its stultifying effects on the moral quality of life but also for the harm done to the character of a culture. "You cannot serve God and mammon" (Matt. 6.24). Social science corroborates these insights. Sociologists, psychologists, and historians argue that given the narrow twentieth-century definition, which equates economics with consumption and its monetary measure, so-called economic progress has not been accompanied by concomitant gains in either moral character or the quality of life.[1]

There is thus more than a little irony in the fact that the American dream of success, a motivational system that every schoolchild is conditioned to accept, so powerfully grips the national psyche. In spite of their religious faith, many Americans appear to believe that wealth and the meaning of life are tied together. Obviously, with wealth comes prestige and power, but too seldom the wisdom to use them ethically. Entrepreneurs often become a parody of themselves, enmeshed in their desire to produce more wealth—at any cost. There are other consequences as well, manifest in America's natural and social ecology, that is, in the exploitation of the earth, now become nothing more than natural resource for the satisfaction of a truncated and scientifically untenable theory of economic growth, and in the stultification of human beingness.

Most frightening of all, our political culture is endangered. Rorty argues that the North American democracies are "under the control of an increasingly greedy and selfish middle class—a class which continually elects cynical demagogues willing to deprive the weak of hope in order to promise tax cuts to their constituents. If this process goes on for another generation, the countries in which it happens will be barbarized" (1991b, 15). We Americans appear to choose our leaders based on a narrow utilitarian calculus. The first question of politics seems to be "What's in it for me?" followed by questions involving the GNP, tax cuts, pay raises, and other narrowly economic concerns. City councils and mayors, representatives and senators, governors and presidents are elected on the promise that they will serve economic ends—even if these ends themselves are at odds. Any moral purpose beyond so-called economic progress is secondary at best. The presidential campaign of 1992 does little to discourage me from thinking that Rorty's judgment, however dire, is accurate.

I note also that Rorty is not alone in his judgment. Many others, both ordinary people and social critics, believe that something has gone wrong in American society. Greed, self-love, and egoism have apparently overrun any countervailing agency that values society and the public interest. Consider the mercenary behavior of our social elites. Members of Congress forced from their positions in disgrace; savings and loan scandals costing hundreds of billions of dollars; insider trading and junk bond scandals on Wall Street; chief executive officers of money-losing and in some cases near-bankrupt corporations drawing millions and even tens of millions of dollars in salary. Does no one think of his or her responsibilities to community, nation, and future generations? Is each and everyone of us consumed by economic self-interest? Is the prototypical member of modern society the "mass person," the spoiled child of history who wants only privileges without obligations?

Caught up in a maelstrom of materialism, voters shirk responsibility for the commonweal. Utilitarian individualism reigns supreme as *the public philosophy*. So absolute is its power, hiding from view anything that does not conform with its description of the world and institutional imperatives, that some doubt our ability to escape. As Hans-Georg Gadamer argues, "If we continue to pursue industrialisation, to think of work only in terms of profit, and to turn our earth into one vast factory as we are doing at the moment, then we threaten the conditions of human life in both the biological sense and in the sense of specific human ideals [love, justice, charity, peace] even to the extreme of self-destruction" (1988, 491). In this context, religious discourse assumes a position of enormous importance, because a metaphor of caring for creation offers *a toehold* for a democratic society to develop solidarity. Political consensus (see chapter 2) is necessary for movement toward sustainability: transformation presupposes the political agreement to do so.[2] The church (see chapter 5) is the obvious place for conversation about an environmental agenda, in the context of sacred stories (see chapter 3) that legitimate caring for creation (see chapter 4), to begin. Religion, as I have argued in the Introduction and throughout this book, offers a genuine possibility for discourse that is an alternative to utilitarian individualism.

Paradoxically, religious discourse is ordinarily associated with *skyhooks*—privileged metaphysical claims—rather than with environmental ethics and green politics. In contrast to toeholds, which give a particular community of believers an opportunity to participate in a public conversation, skyhooks appear to close rather than open conversation, since they are assertions of ultimate knowledge. Religious wars, if nothing else, testify to the gravity of such beliefs, for the existence of rival claims to ultimate knowledge is a matter of utmost consequence. For the metaphysical realist the notion of multiple, socially defined realities is a contradiction in terms. Clearly, metaphysical claims to knowledge, such as the belief that Our Father in Heaven is the one God who is there, are important to the faithful.[3] One anonymous reader of a manuscript version of this book correctly reminded me that the power of the sacred makes religions tick. Not only the keenest students of religion, such as Mircea Eliade and William James, but history itself confirms the importance of the numinous, the transcendent, the wholly other, to believers.

My argument, whatever the appearance created by my claim that religion offers toeholds to deal with ecological crisis, is not that claims to ultimate knowledge are unimportant. *Without a sacred canopy of ultimate belief that frames the world in which the believer lives, there can be no toeholds.* Religion, more than anything else, confirms for believers their place in a meaningful cosmos, a telos that is an ultimate guide, a reality that legitimates existence. Indeed, my pragmatism presupposes that the faithful have a sure and certain belief in their religion. I also acknowledge the diversity of religious belief in American society. My approach celebrates and depends on this diversity, rather than reduces it to an operationally common denominator, that is, some bland, homogenized agreement on a secular environmental agenda.

My point in emphasizing the political implications of religiously derived commitments to care for creation is not to undercut claims to ultimate knowledge. As chapter 4 shows, faith in the God who is there, as revealed in the Bible, enables conservatives to take meaningful actions in response to ecocrisis. Similarly, postpatriarchal Christians believe that it is true (that it is actually the case, or that the reality is such) that Christians cannot care for creation until they overcome the dichotomization of God and Gaia, spirit and matter, history and nature, and man and woman inherent in the fusion of the Old Testament with Greek philosophy. I assume that ultimate commitments are germane to believers and congregations of believers. *Politically considered*, however, insofar as American citizens are going to deal with ecocrisis, the issue is not one of sorting through the relative adequacies and inadequacies of competing claims to ultimate knowledge. Instead, the issue we confront *as citizens* is one of solidarity. Caring for creation, whatever its metaphysical implications, can be a politically effective metaphor, because it cuts across the continuum of religious belief—a spectrum that encompasses 90 percent of the populace.

To make the same point in a different way, if religious discourse was only of metaphysical consequence, then a metaphor of caring for creation would not be politically useful. But metaphysical consensus (alternatively, a master narrative) is not required for a pluralistic society to take political action. Indeed, by definition, the citizens who make up a pluralistic society have a diversity of ultimate commitments. Yet the diversity of claims to sure and certain knowledge enables toeholds for citizens to achieve solidarity on an environmental agenda. Religiously inspired and democratically articulated preferences are the basis of green politics but not of a theocratic state. Since democracy, as virtually everyone agrees, is a first-order value, then a green or sustainable society will necessarily preserve such democratic freedoms as speech, assembly, press, and suffrage. Crucially, the diversity of religious constituencies works in favor of ensuring that a green society will be democratic. Every tradition of faith has a vested interest in protecting the political freedoms that enable its own existence. Solidarity on caring for creation is fully consistent with liberal-democratic life.

The genius of democratic life is not majority rule. It is that policy issues are tested in the give and take of democratic process—that is, conversation and debate. Granted, diverse claims to ultimate knowledge will sometimes cause difficulty on certain environmental issues, such as population control. But the diversity of religious belief promises to spill discourse outside the banks in which the prevailing public philosophy channels it—that is, the idea that the social good equates to the sum of narrowly defined, individual economic goods, primarily monetary income. As Bellah argues, "Our problems today are not just political. They are moral and have to do with the meaning of life. We have assumed that as long as economic growth continued, we could leave all else to the private sphere. Now that economic growth is faltering and the moral ecology on which we have tacitly depended is in disarray, we are beginning to understand that our common life requires more than an exclusive concern for material accumulation" (Bellah et al. 1985, 295).

Culture as Text

The Bible is an exemplary text that presents a number of interpretive possibilities. (There are also a number of possibilities that are outside the biblical tradition. Because most voters are Christians, however, the biblical tradition is central.) This array of interpretive possibilities makes the Bible protean, able to take on new meaning within changing circumstances. George Lindbeck argues that "*the vitality of Western societies may well depend in the long run on the culture-forming power of the biblical outlook in its intratextual, untranslatable specificity*" (1984, 134, my emphasis). The cogency of Lindbeck's thesis, as well as the argument advanced in this book, depends on human behavior. The issue, in other words, remains open.

The established structure of Enlightenment ideology and institutions controlling American society make change difficult. The world is there, ready-made, seemingly complete and self-sufficient in and of itself—a specious present that obscures the potential for change. Transformation comes through conversation—through discourse that challenges the status quo, the legitimating narratives, taxonomies, and ideologies that enable culture. Our corporate life is governed by a narrowly utilitarian discourse. Its basic premises are that humans control the earth through the power of science, that they are above the Creation, that modern culture is an unqualified success, and that things will get better by fine-tuning the industrial growth paradigm. What is ignored in this modern story is the origin myth that legitimates it.

Utilitarian individualism itself reflects the vagaries of history. The rhetoric of Modernism is persuasive because its master architect, Francis Bacon, was able to trade on the legitimating narrative of Judeo-Christianity. Which is to confirm, again, the power of the sacred canopy: we belong to language. As Bacon intuitively sensed, Judeo-Christianity provides a basic telic structure apart from which our lives (religionist and atheist alike) are almost incomprehensible. Bacon was a man who saw through time and so straddled two ages: the Medieval Age, focused on the Kingdom of Heaven, which had settled into lethargy and poverty, and the Modern Age. He heralded the possibility of harnessing the power of physics and applied technology to transform a fallen world (caused by original sin). Bacon, in concert with Newton, believed that God was the master designer of the cosmic engine: science would restore humankind to its prelapsarian dominion over nature. *Man* was to rule over the ecomachine as its master and possessor. In this way science could lead humankind to the New Jerusalem. The modern project became the domination of nature.[4] As Mark Backman suggests, "The power of rhetoric resides in its ability to discover, in the materials at hand, language and modes of interpretation suitable to immediate circumstances" (1987, xxi).

Ironically, once the modern paradigm was established, organized religion became increasingly irrelevant to the cultural mainstream, since science and technology were to restore what the Fall had put asunder. Religious discourse refocused on the supernatural, especially salvation. Utilitarian individualism became the lingua franca of the common quest, driving the West in a relentless upward spiral of materialism. Today the pursuit of wealth has become a parody of itself, distorting both social ecology, that is, our personal and communal lives, and natural ecology. And this is paradoxical, because the public philosophy makes the hold of Americans on their private beliefs ever more tenuous. Materialism, in a phrase, violates our religious affirmations: our beliefs that the Creation is good, that we have a responsibility not to impair the Creation, and that our neighbors are our sisters and brothers.

Economists have become the high priests of American society because economic theory creates the illusion that it has objective and nonpartisan knowledge that does not infringe on our diverse claims to ultimate knowledge. The truth is a different matter. In the first place, as Ekins and others argue, economic theorizing and the decision making it engenders "should be recognized as the political act that making decisions about society always really is" (1992, 37). Further, economic theory should not be thought of as some unchanging, absolute knowledge that mirrors the way things really are. It is more a matter of culturally conditioned storytelling than objective rationality. Clearly, economic theory can be reformulated to incorporate a more inclusive social ecology and a greater grasp of natural ecology. It is not set in concrete.

As Rorty explains, we can abandon any appeal to reason "conceived as a transcultural human ability to correspond to reality" (1991b, 28); this notion has lost its ability to help us change our behavior. Instead we can begin to think in terms of the importance of metaphor. The metaphor of caring for creation is literally an instrument of social transformation; it is an instrument of moral and intellectual growth. I do not mean to imply that religious language about creation sets down absolute guidelines for moral progress, even though I acknowledge that many if not most of the faithful remain secure in their claims to ultimate knowledge. There is neither inconsistency nor contradiction here. Caring for creation is not a theological rule but an imaginative paradigm that might prove useful for a culture undergoing ecocrisis. So viewed, religious discourse expresses not only private beliefs (skyhooks) but helps us articulate our collective preferences as citizens.

One way to comprehend the potency of the metaphor of caring for creation is to emphasize its roots in Judeo-Christian origin myths. Origin myths, to put it bluntly, have culture-forming power. They possess both legitimacy through the prestige of origin and potency through the evocation of sentiment (emotion). As discussed in chapter 2, religious story or myth provides a sacred canopy that legitimates existence. Origin myths, as Geertz (1973) puts it, are both models for and models of reality. Myths are, argues Bruce Lincoln, "discursive act[s] through which actors evoke the sentiments out of which society is actively constructed" (1989, 25). This is the awesome power of sacred story. The Great Code is protean, in part because its origin myths can be reinterpreted in today's circumstances. Through reinterpretation—especially through religious narrative that emphasizes caring for creation—feelings of care (*sorge*, as some have called it) that have been overwhelmed by the passions of materialism are evoked.

Sociobiology (see chapter 2) helps us realize that myths of origin are essential to the modification of selfish meme complexes, restoring potency to religion. For all the controversy that it has created, sociobiology does not threaten but ennobles human dignity. From a sociobiological perspective, humans are distinctive because of our biological underdetermination and cultural overdetermination. The uniqueness of our social behavior "pivots," E. O. Wilson argues, on our "use of language, which is itself unique" ([1975] 1980, 280). Religious discourse is a unique kind of language, since it is the source of the sacred canopy. Memes, the replicators that make culture possible, are carried by language. Almost like genetic selection, the cultural selection process favors memes that selfishly exploit their environment. Utilitarianism is the linguistic means by which the industrial growth paradigm (a co-adapted meme complex) perpetuates itself. Language speaks, Heidegger says (1971). To what extent, Gadamer wonders, do we find ourselves in our present predicament "because of the baleful influence of language"? (1988, 491). Yet does not the industrial growth paradigm—modern society—epitomize this speaking that is language? Its organization, from the infrastructure of transportation systems and water resource projects to the institutional superstructure of public bureaucracies and private corporations, is dedicated to one narrow purpose: maximization of the production-consumption cycle. Discussion of our common purpose is monopolized by the constraints of the Dominant Social Matrix—which is to acknowledge that memes perpetuate themselves by driving off their competitors.

Although we are caught up in institutions that circumscribe our economic and political lives, we can free ourselves. The human gift to imagine the future, Dawkins suggests, frees people from the power of selfish replicators—which is to say that the possibilities of interpretation explain the demonstrated ability of the Great Code to renew itself time and again. As Wilson argues, "The fully symbolic quality of the words and the sophistication of the grammar permit the creation of messages that are potentially infinite in number" ([1975] 1980, 280). We can deny the selfish meme complexes that overdetermine us all. As Gadamer puts the point, "Language is not [only] its elaborate conventionalism, nor the burden of pre-schematization with which it loads us, but the generative and creative power unceasingly to make this whole fluid" (1988, 498). Language opens up the possibility of freedom, the freedom of "speaking oneself" and of "allowing oneself to be spoken" (498).

The future is unimaginable apart from biblical language, if for no reasons other than our basic idea of history as a meaningful sequence of events and our idea of the person as a free agent somehow in control of its destiny. These are among the most basic and therefore controlling memes of our culture. To set religion in a sociobiologic context in no way circumscribes the power of religion to change human behavior. According to Wilson, the genetic influence on mental development, "even when very strong, does not destroy free will. In fact the opposite is the case: by acting on culture through the epigenetic rules, the genes create and sustain the capacity for conscious choice and decision" (1983, 182). On my reading, Wilson is ambivalent about religion, recognizing it on the one hand as a profound and strong, perhaps the strongest, influence on human behavior (1978, 169), and on the other hand as "an archaic procedure [that] just might, by fantastic good fortune, lead in the most direct and untroubled manner to a stable and wholly benevolent world" (1983, 184). More likely, in Wilson's opinion but not my own, religion will "perpetuate conflict and continue to drag humanity relentlessly along what is at best a tortuous and agonizing path" (Wilson and Lumsden 1983, 184). Living in a time of ecological crisis *and* in a democratic society, I see no alternative to reaching an environmental ethic that grows out of religious truth claims. Time, as Wilson emphasizes (1992), is of the essence.

Consistent with the idea that cultures are grounded in and sustained by their classic or exemplary texts, my project has been to read the texts of ecology and politics against the Great Code specifically and religious discourse more generally. "Written texts," Tracy suggests, "seem to provide stability for literate cultures. At the same time, written texts are exposed to great instability when intellectual and moral crisis occurs" (1987, 11). We live in and through our legitimating narratives; so viewed, the Bible is more verb than noun. If we are to survive a time of crisis and emerge, renewed, on the other side, then as a literate culture we must return for inspiration to our exemplary texts. The interpretive question is this: Does the Great Code illuminate the present and show the possibility of transcendence? Or does it stand mute and barren, more fossilized remnant of an age gone by than vehicle of conveyance to a new era? Are we prisoners of a failed narrative that has outlived its relevance to life, incapable of birthing a new age, or does that narrative offer the interpretive possibility for a new era?

Interpreting the Great Code: Idealism, Psychologism, and Pragmatism

A culture necessarily stands in a vital relation to its classics, for these texts are woven into the language of life—the stories people tell that direct their behavior and give their lives meaning. Tracy suggests that "the historical memory of a people is the principal carrier of the history of effects of the classic texts, persons, events, symbols, rituals of that people. It is also true that a loss of those memories, in either an individual or a communal sense, can be fatal to participation in a culture. For without them we cannot act. To become socialized by learning one's native language is to give life to all those carriers of meaning and action that are a tradition" (1987, 36). The crucial question is how to interpret an exemplary text—one that has shown the ability to be renewed through interpretation. The answer is that there are *at least* three possibilities (and these are not necessarily exclusive): idealism, psychologism, and pragmatism.

Idealism

For idealists the interpretive process permits alternative interpretations of texts within a larger framework that restrains diversity. "Polysemous meaning," as Frye has it, "is the development of a single dialectical process, like the process described in Hegel's *Phenomenology*" (1981, 222). So conceived, the Bible takes on new meanings through the historical progression of Western culture—the continuing articulation of a single dialectic. Frye's concern is with the Bible as a verbal structure that appears inexhaustible, capable of constant renewal through interpretations that find relevance to life (unlike, for example, detective stories or romance novels), but in a way that remains true to a continuous tradition. Frye's view of interpretation, then, is neo-Hegelian: the "Absolute" does not exist for Frye apart from the ongoing process of interpretation.

> The hero of Hegel's philosophical quest is the concept (*Begriff*), which, like Ulysses in the *Odyssey*, appears first in an unrecognized and almost invisible guise as the intermediary between subject and object, and ends by taking over the whole show, undisputed master of the house of being. But this "concept" can hardly exist apart from its own verbal formulation. . . . What Hegel means by dialectic is not anything reducible

to a patented formula, like the "thesis-antithesis-synthesis" one so often attached to him, nor can it be anything predictive. It is a much more complex operation. [222]

Psychologism

In contrast to idealism, interpreters may opt for the psychologism of Freud in their interpretation of interpretation.[5] Following Vico, for example, Harold Bloom attempts to escape idealism "by returning authority to our historical births, and defining authority as poetic wisdom or the imagination" (1975, 67). For Platonic metaphors and the rational mind of idealism he substitutes the unconscious mind and the psychoanalytic metaphors of Freud. Truth, according to Bloom, is not (transcendent) form articulating itself rationally through time (Hegelian or dialectical interpretation); rather, it is created through the workings of the unconscious, the creative imagination. Bloom rejects the Heideggerian-Wittgensteinian notion that language speaks, that we belong to language, substituting a Freudian notion that "*belatedness* or the fear of time's revenges is the true dungeon for the imagination, rather than the prison-house of language as posited by Nietzsche, Heidegger and their heirs" (68). Yet surely Bloom must admit, by recognizing the contingency of his own interpretive gesture, the possibility of alternative readings of Freud. Rorty's reading in *Contingency, Irony, and Solidarity,* for example, uses Freud's texts to undergird his contention that the self is a historical contingency. But the possibility of interpretation is not tied, as Rorty recognizes, to Freudian narratives. The value of Freud, according to Rorty, is that he "lets us see the moral consciousness as historically conditioned, a product as much of time and chance as of political or aesthetic consciousness" (1989, 30).

Freudian theories of interpretation, by this logic, are possible but not necessary, being themselves only contingencies. No one language, including psychoanalytic discourse, exhausts the possibility of descriptions. Bloom transforms the language of psychoanalysis from its role as a therapeutic paradigm to a theory of reading and culture (from a patient on the couch to individuals suffering from the "anxiety of influence," that is, fear of the loss of freedom, of control over destiny). But Bloom's use of Freudian narrative is privileged—it is insulated or set off from the possibility of further interpretation. His account of Freud thereafter influences (indeed, overdetermines) his other reading (and more particularly his theory of reading).

The value of Bloom's interpretive gesture lies in his notion of the *anxiety of influence,* for it illuminates our dilemma. Mired in a worsening ecocrisis, Americans are caught between a failed past and a future powerless to be born. The modern (bourgeois) mind seeks to escape ecocrisis through the legitimating narratives of utilitarian individualism, through engineering, through managing planet Earth. For the modernist, talk of sustainability promotes the fear of savagery, of primitivism. In contrast, the environmentalist fears that the Great Code itself has brought ecocrisis upon us. How can religion be any part of the solution? Even religionists, as many note, fail to recognize that the Great Code can engender a postmodern mythos: they fear an uncertain future. The temptation is to hold onto the past, the tradition. The consequence is sociocultural paralysis: the industrial growth society drifts on toward oblivion.

Pragmatism

Where I depart from Bloom's psychologism and Frye's idealism is on the notion of language itself. Bloom's rejection of the idealization of interpretation inherent in Frye's position is defensible. The problem with idealism is that the reader (critic or student) becomes "a cultural assimilator who *thinks* because he [or she] has *joined* a larger body of thought" (Bloom 1975, 30). Yet novel response to changing circumstances of existence entails the notion of freedom, of volition not totally constrained by tradition. "Freedom," Bloom maintains, "is *the change*, however slight, that any genuine single consciousness brings about in the order of literature simply by joining the simultaneity of such order" (30, my emphasis). Bloom and Frye thus part ways on how to read the Great Code, since interpretation for Bloom "now seems invested in the interplay of repetition and discontinuity, and needs a very different sense [than idealism] of what our stance is in regard to literary tradition" (30).

The problem with Bloom's psychologism is that, though he denies it, he implicitly endorses a sociolinguistic approach to language. In fact, by being human he cannot (nor can any human being) escape it. Bloom belongs, as his privileging of Freudian texts confirms, to language. According to Bloom, language (textual tradition) is the father. In interpretation "the father is met in combat, and fought to at least a stand-off, if not quite to a separate peace" (1975, 80). Read here "father = language = authority = origin"; we belong to our father = language = authority = origin. So far, so good. But Bloom's use of Freud, though ingenious, is deceptive because it conceals a more fundamental truth about language and interpretation. Authority is in language, but the most basic texts are those that relate *myths of origin*, not psychological theory. This is the crucial point. To put the point even more directly: we read the Bible to find out why we even read at all.[6] The Yahwist, not Freud, is primordial, crucial to the imaginative tradition of the West. (Bloom confirms the cogency of this notion by arguing that the primal author J has an influence equal to, indeed, greater than, Shakespeare's. See Bloom 1989 and 1990 on this point.)

There is, however, as Rorty argues, a contingency to all textual traditions—including biblical and Freudian ones.[7] We forget this truth because the world seems to confirm that language names things the way things are. But "the world does not speak. Only we do. The world can, once we have programmed ourselves with a language, cause us to hold beliefs. But it cannot propose a language for us to speak. Only other humans can do that" (1989, 6). This means that there is a contingency to culture: no culture is "set in concrete"; it must always remain open to the play of language. Neither is any theory of interpretation absolute. For the pragmatist the questions are these: Is one interpretation (or theory of interpretation) more useful than another? And what texts, specifically, are to be reinterpreted?

Other interpretive theories, I hasten to add, are possible, including those derived from biblical hermeneutics. As a sociolinguist, attempting to describe the process whereby religious discourse might move our society toward sustainability, I stand outside any one tradition of faith and its theory of interpretation. I do not seek to deprive any religion of its spiritual potency, of its call to the faithful, but rather I seek to show how alternative readings of the Great Code, whether conservative or radical, are useful in creating solidarity. If my critics charge that this is "lowest common denominator religion," then so be it. But please note that I do not offer my pragmatic account of interpretation *in lieu of a faithful reading of the Great Code.* My pragmatic frame is just that: a frame that allows us to conceptualize diverse religious belief systems as converging on a center of caring for creation.

As Umberto Eco argues, not all interpretations are permitted. "A text is a place where the irreducible polysemy of symbols is in fact reduced because in a text symbols are anchored to their context. . . . Texts are the human way to reduce the world to a manageable format, open to an intersubjective interpretive discourse" (1990, 21). Frye glimpses this truth in noting that the literal "Bible of myth and metaphor . . . combines with its opposite, or secular knowledge, the world of history and concept which lies outside the Bible, but to which the Bible continually points. It points to it because it grows out of that world, not because it regards it as establishing criteria for itself" (1981, 228). Which is to say, again, that the Bible offers the West, at this perilous time, a sacred canopy. "The structure of secular knowledge," Frye continues, "so far as it bears on the Bible, is not only a rooting of the Bible in its human context, but a manifesting of the human struggle to unify its world" (228). Religion alone cannot see us through to sustainability. But without it we have no future, since it is the wellspring of our cultural distinctiveness—the Western world in which we live and move and have our being. And it is also, as sociobiologists like Wilson and cultural anthropologists like Geertz make theoretically clear, and as believers across the entire continuum of faith know immediately through the power of the Spirit in their own lives, the way that human beings are empowered. As I noted in the last chapter, hope for a better world rests on something more than utility; it depends mightily on faith.

Americans, secularist and fundamentalist alike, are set in a Judeo-Christian context. Although frameworks other than religious ones can lead to a caring for creation, like Aldo Leopold's land ethic, it is difficult to imagine, especially in light of the paradox of environmentalism, that any other framework might work so effectively to create solidarity. Robert Paehlke, for example, persuasively argues that ecological narrative empowers environmentalism. Many of the keenest students of ecology, such as Neil Evernden and Robert McIntosh, argue *against* the idea that ecology has been useful in dealing with the fundamental problem, the discourse of power that reduces the creation to standing reserve. My argument is that religious discourse might actually empower ecological narrative such that it could make a difference in ecological outcomes—namely, by grounding an environmental ethic to care for creation.

Discourse and the Construction of Society

Living inside the definition that "*man* is the rational animal," Americans have lost sight of the psychological, the affective side of human beingness. The truth is that we cannot grasp ourselves as distinctively human apart from recognizing the reality of the subcortical brain that conditions our emotions. Wilson argues that the hypothalamic-limbic complex of any social species, including our own, "has been programmed to perform as if it knows . . . that its underlying genes will be proliferated maximally only if it orchestrates behavioral responses that bring into play an efficient mixture of personal survival, reproduction, and altruism" ([1975] 1980, 3–4). Lincoln puts the point more directly in relation to religious discourse: "that which either holds society together or takes it apart is sentiment, and the chief instrument with which such sentiment may be aroused, manipulated, and rendered dormant is discourse" (1989, 11). Religion, viewed from outside the framework of any particular belief, is the primordial expression of human emotion.

Society is an artifice, a construction. Like a building, the facade deceives us, concealing the supporting structure beneath the surface. In a totalitarian, fascist, or aristocratic society that skeleton (legitimating narrative) is frozen in place and often maintained by force. If we do not respond democratically to ecocrisis, administrative despotism, the totally managed society, is likely. But in a democracy the possibility of changing the legitimating narrative is comparatively open, although change is always difficult. Religious discourse, I have been arguing, is the most likely way for the supporting narrative of our political economy—a self-perpetuating structure that drives off competitors—to reweave itself. Lincoln argues that "society in its essence consists not of impersonal structures, but of human beings who feel or come to feel more affinity for, than estrangement from, one another. This being so, society is reconstructed in some measure whenever sentiments emerge that draw together those who previously felt separated or separate those who previously felt commonality" (1989, 173). Through the play of language, through discourse that stresses caring for creation, solidarity on an ecologically sustainable society can be created. But solidarity requires that sentiment be evoked, that the psyche be mobilized.

"Greens" sometimes claim that postmodern religion will conform to or be inspired by an ecophilosophical narrative such as deep ecology or ecofeminism. There is no reason to believe that critical reason can persuade many Americans to care for creation. Indeed, as noted, *even if environmental philosophy achieved theoretical unity, nothing would change*. Cultural transformation presupposes solidarity on the broad outlines (adjusted to fit specific regions) of an environmental agenda. These changes, in turn, must be institutionalized. As discussed in the Introduction, it is our very means of moving ourselves on the surface of the earth that contributes to the global greenhouse, just as our demands for cheap plywood and beef threatens biodiversity. Until citizens change their political behavior, no claim that secular environmental ethics is useful can be legitimated. This means that the controlling memes (the story) of industrial culture must be reshaped. And this can only be done through discourse that is widely communicated, ideologically persuasive, and emotionally evocative. Professional environmental ethics is not up to the job—it is too technical in nature and it lacks standing across society. Neither does it deal in origin myths.

Similarly, there is no reason to think that scientific discourse, in particular ecology, can usher in a sustainable society. Ecology will be involved, since discourse that hopes to be useful must be consistent with whatever it is that nature permits. (For example, regardless of how the behavior is motivated and legitimated, nature does not allow megatonnages of CO_2 from a hydrocarbon economy to be pumped into the atmosphere without consequence.) A critical examination of conservation ideology reveals that it has been and continues to be dominated by the very philosophy of utilitarian individualism that is undercutting natural ecology. In spite of Anna Bramwell's fears (see chapter 2), there is no reason to think that ecology can provide psychologically potent and politically effective narratives. McIntosh concludes his study of ecology with the judgment that, whatever the social expectation, ecology remains reluctant, unlikely to serve as a guide for the perplexed. Professional ecologists themselves argue that if sustainability is to be achieved, ecology must be empowered, by societal factors that are outside the domain of ecology proper, to integrate itself with the policy-making process (Levin 1992). But, as Evernden argues, ecology has not been permitted to participate in a genuine conversation about the heart of ecocrisis; it has only served as a bandage to mask the unseemly appearances of environmental wounds. "Ecology," he writes, "as a science based on the implicit expectation of prediction and control [of nature for human benefit], fits all too well into the civilizational destiny, particularly in its 'Imperial' form" (1992a, 77).

Somewhat like Sayre, Hargrove (1992) argues that most of us make value judgments not by theorizing philosophically about values but by referring to a repository of existing cultural traditions, such as judgments of beauty. This is consistent with a sociolinguistic perspective on religious discourse. People decide to protect nature or assume their obligations toward future generations not because they are underpinned either by philosophical theories or by scientific fact but because such decisions express ethical obligations articulated in the context of traditional stories. *And for the vast majority of North Americans, the one vernacular tradition outside the language of utilitarian individualism is religious discourse.* Every faith, including and especially Judeo-Christianity, can articulate a compelling sacred story, based on the metaphor of caring for creation, to treat nature with respect. In Hargrove's (1992) terminology, religion is the primary source of weak anthropocentric intrinsic value, that is, those descriptions of and deliberations about natural ecology that trump claims of instrumental value.

Because we live in a secular society in which religion has become increasingly privatized, the claim to have found potential for a politically empowered environmental ethic in the Great Code seems paradoxical. Yet the situation is not surprising: for the borders of society by which groupings are created are never so firm as proponents of the status quo believe. The situation is rather the opposite. Lincoln suggests that "there always exist potential bases for associating and disassociating one's self and one's group from others" (1989, 10). Liberal pragmatists have simply assumed that religion promoted dissolution rather than association. And environmentalists have simply assumed that religion was the problem, or at least a major part of it. What is essential, as Lincoln emphasizes, is the recognition that "the vast majority of social sentiments are ambivalent mixes in which potential sources of affinity are (partially and perhaps temporarily) overlooked or suppressed in the interests of establishing a clear social border" (10).

Achieving sustainability in a society that also aims to preserve its democratic values necessarily means that diverse religious groups, acting out the implications of their claims to ultimate knowledge, will come together in a consensus on an environmental agenda that encompasses them all. Creating a sustainable society presupposes such solidarity, a new synthesis. Society, Lincoln argues, is a thing "con-structed, put together (from the Greek compound verb, *syn-tithēmi*, to put or place together)" (1989, 11). Creating solidarity on caring for creation does not, however, depend on a master narrative. No one rationale, no one ecotheology, is presupposed. Local churches offer a rich and diverse array of myth-ritual-symbol complexes that evoke sentiment and offer reasons to care for creation.

All of this means, finally, however Lynn White's argument (see chapter 1) was once read by environmentalists, that a new reading is in order. White himself was a religious (Presbyterian) layperson. "The Historical Roots of Our Ecologic Crisis" can and perhaps should be read as being in the Judeo-Christian tradition of prophetic self-criticism.[8] White was less counseling Judeo-Christians to abandon their faith in favor of Native American or Eastern religions as challenging them to renew their faith in the tradition of those who cared for creation. While most environmentalists know the article he wrote in 1967, many probably have not read a second piece, written in 1973, entitled "Continuing the Conversation." This article clarifies the remark (found at the end of the first essay) that "since the roots of our trouble are so largely religious, the remedy must also be essentially religious" (1973, 30). In this amplification White mentions at least twice the potential for change inherent in Judeo-Christianity. "In its doctrine of the Holy Spirit, Christianity fortunately makes provision for continuing revelation. Or, to phrase the matter in a more orthodox way, it recognizes the progressive unfolding of truths inherent in an original deposit of revelation. The Christian wants to know what Scripture says to him [or her] about a puzzling problem" (60). Then, a bit later (with a nice twist of biological metaphor) White goes on to say this:

> In every complex religious tradition there are recessive genes which in new circumstances may become dominant genes. In my 1967 discussion I referred to St. Francis's abortive challenge to the anthropocentric concept of God's world. Scattered through the Bible, but especially the Old Testament, there are passages that can be read as sustaining the notion of a spiritual democracy of all creatures. The point is that historically they seem seldom or never to have been so interpreted. This should not inhibit anyone from taking a fresh look at them. [61]

The Great Code as Story

Described in sociolinguistic terms, the power to transform ourselves belongs to language, an open-ended process of meaning transforming itself. While myths of origin can become transcendental signifieds, nearly impervious to change, there is no historical inevitability in this: sacred stories remain subject to the play of language. *Put in religious terms, sacred story remains open to continuing revelation.* More than anywhere else, this play of language can occur in the church.[9] The metaphor of caring for creation is the play of language: religion in a time of ecological crisis. Of course, there are limits to interpretation. As discussed in chapter 4, *conservative creation stories* are the most restrictive in interpretation; still, they provide ample grounds to care for creation. And there an array of *moderate and liberal interpretative possibilities* exists. Within the Judeo-Christian spectrum, *radical creation stories* go the farthest, making accommodations to scientific discourse that conservatives find anathema. *Alternative creation stories* go even farther, beyond the pale of Judeo-Christian myths of origin.

Environmentalists and religious believers, I have argued, are coming together in a new era of green politics. There is a potential for solidarity on sustainability that does not transcend but nevertheless withstands differences on other issues. Charlene Spretnak contends that a "religion-based movement for social change is beginning to flourish that is completely in keeping with Green principles" (1986, 65). Even more, I believe, is portended by the convergence of environmentalism with Judeo-Christian narrative. As nonsensical as this may sound to a secular environmentalist, outside a Judeo-Christian narrative tradition this very discussion is impossible. Gadamer argues that "the effective-historical consciousness is so radically finite that our whole being, achieved in the totality of our destiny, inevitably transcends its knowledge of itself" (1988, xxii). How storytelling culture-dwellers transform themselves is not mysterious. For the Great Code contains in both of its testaments the seeds of change, and this is so for several reasons.

One is that our sense of time is incomprehensible apart from a Judeo-Christian narrative tradition. The Old Testament literally brought into being the sense of temporal process as meaningful, as going somewhere. And yet the process is mysterious, for the future is open. As Marty puts it, "The future remains horizontal, which means visible as a presence but utterly unknowable in detail. . . . [Human beings] act on the basis of awareness of who they are, of past actions, of what they can get from story, from history. They add to the story, and thus provide grist for interpretation by the philosophers, or the writing of new chapters by later historians" (1989b, 21).

Another springs from the notion that Judeo-Christianity is a parabolic rather than mythic religion—a point that can be stated in a variety of ways. One is to note that the truth value of the propositions asserted by Judeo-Christians remains important to them. Although the Western idea of knowledge is customarily thought of as originating with the Greeks, the hebraic prophets were there several hundred years earlier. They were the first to dissociate themselves from the restrictive confines of culture, that is, the conventional wisdom, and associate truth with something beyond that which was immediately before them. As Schneidau succinctly summarizes it, "The Bible, read objectively, tells us why we read objectively" (1976, 21). The consequence is the Western drive to criticism and interpretation. "What unites Western culture in all its phases," Schneidau continues, "tying in with the ambivalence that produces the continuity of change, is a series of demythologizings and consequent 'losses of faith'—some gradual, some traumatic. Nothing is so characteristic of our traditions, with the result that we can say more truly of Western culture than of almost anything else, *plus ça change, plus c'est la même chose*" (14).

Finally, we might recall that the Judeo-Christian tradition has two testaments, the New Testament adding the parables of Jesus to the prophecy and parables of the Old Testament. According to John Crossman, parables "are stories which shatter the deep structure of our accepted world and thereby render clear and evident to us the relativity of story itself. They remove our defenses and make us vulnerable to God. It is only in such experiences that God can touch us, and only in such moments does the Kingdom of God arrive" (1988, 100). Viewed in this way, Judeo-Christianity is a parabolic rather than mythic religion, "a religion that continually and deliberately subverts final words about 'reality' and thereby introduces the possibility of transcendence" (Crossman 1988, 105).

Marty argues that all traditions, including Judeo-Christianity, begin radically. Just as the Old Testament appropriates the origin myths of its precursors,

Table 8 An Overview of the Interpretive Process

- Individuals learn their story in community; by learning their story (through socialization), people learn to be human.
- Story traditions sometimes develop anomalies, either through internal inconsistency or by failing to address the problems of life.
- Human beings cannot, however problematic their lives, invent themselves ex nihilo; the old story must be preserved and carefully retold, with the crucial events and consequences noted.
- From the old story there is an imaginative alteration in which the community remains faithful to its story while simultaneously envisioning an alternative future—a new use for it.
- Through an interpretive encounter with story (that is, with the canon) individuals learn the world—that is, they are able to imagine and create a future.
- Language is performative—that is, myths of origin have culture-forming power because they are evocative of sentiment and therefore motivational and intellectually legitimate and therefore plausible.

Source: Adapted from Bloom 1975, Marty 1989b, Lindbeck 1984, Lincoln 1989, and Rorty 1991b

so Christianity took over the Pentateuch. "Each taking over alters what another community had made of a story, but the appropriators need the earlier story as they add new interpretations and events" (1989b, 16). The process by which a culture, or more accurately, the diverse subgroups that constitute a culture, changes its story is schematically outlined here (table 8).

The reinterpretation of sacred story in a time of ecocrisis does not sacrifice scientific judgment on a holy altar of religious truth. *Caring for creation is consistent with ecology.* We should recall that any science, including ecology, ultimately finds its purpose in a larger cultural context: it does not define that context. Ecologists themselves are acutely aware of this (Levin 1992, 213). Religious discourse can be consistent with scientific discourse, that is, ecologically plausible, but it can also evoke sentiment. So interpreted, religious discourse that articulates a metaphor of caring for creation is fully compatible with our controlling memes—on personality and on time. These memes of person and history distinctively mark Western culture.

Conclusion: Church, Person, Politics

Our culture is caught between a failed past and a future powerless to be born. Religious discourse has been privatized, pushed by utilitarianism to the edge of our cultural conversation. The utilitarian glorification of the individual as Homo oeconomicus has enabled the new industrial state—a political apparatus that, ironically, undercuts the Enlightenment narrative of individual freedom. Further, the power of science, whatever its potential contribution to a sustainable society, has been appropriated by this secular leviathan, largely for the purposes of a narrowly defined and scientifically outmoded theory of economic growth. We live in an era of controlled capitalism where ordinary people, deprived of *effective participation* in a political community, find themselves threatened by statism.

Yet change is possible. We the people can lead the movement for change. Social innovation presupposes:

- that an inclusive array of different groups comprising society engage actively in conversation about their collective purposes;
- that this conversation be persuasive, that is, intellectually substantive and capable of winning the assent and commitment of individuals who constitute different groups; and
- that this conversation be capable of evoking sentiment, that is, of moving people psychologically. Religious narrative, the Great Code, is the only form of discourse that presently meets this bill of particulars.

Even the secular environmentalist needs the biblical tradition, for our culture is incomprehensible apart from its myths of origin. Schneidau argues that "language is the great original 'system of difference' which reaches and brings together all members of a culture; myth, a special kind of language, supports many other 'systems of difference' (kinship systems, geographics) which allow the culture to elaborate itself" (1976, 51). Because myth is language, then myth, as Eric Gould so brilliantly discusses, "is both hypothesis and compromise. Its meaning is perpetually open and universal only because once the absence of a final meaning is recognized, the gap itself demands interpretation which, in turn, must go on and on, for language is nothing if it is not a system of open meaning" (1981, 6). The metaphor of caring for creation, springing out of our distinctive Judeo-Christian myth of origin, is the last, best chance for Americans to fashion a sense of community beyond the market and construct a sustainable society.

So redescribed, the culture-forming potential of the Great Code, the source of our own Western myth of origin, can be grasped inclusively, that is, in a way that all faithful people can appreciate. Inevitably, because language is open, not everyone will find my description of sacred story consistent with theirs. But the aim of my account of the role of religion in a time of ecocrisis has not been to favor any one claim to ultimate knowledge (skyhooks) but rather to empower them all in a way that leads to effective public discourse (toeholds). The interpretive question I pose to divergent traditions of faith is not that they justify their metaphysical claims. Rather, I ask the many believers who form congregations and the many congregations that form denominations to articulate their own religiously grounded environmental ethic that cares for the Creation. To fail to do so is for a religious conservative a sin; for a liberal it is bad faith. This discourse can lead us toward sustainability. As I have argued, any society that is not sustainable is a priori self-defeating and therefore not a good society. Obviously, given this chapter, I favor a pragmatic theory of interpretation. But I do not close off either the possibility or the legitimacy of other theories. My redescription of faithful people as *Homo narrans* does not preclude other descriptions, such as the religious conservative's belief that man is made imago dei. I claim only that a pragmatic perspective is a useful way to explore the possibility that a wide diversity of traditions of faith might contribute to a public conversation that addresses ecocrisis.

The array of interpretive techniques and possibilities is rich. A religious conservative, such as Francis Schaeffer, proceeds very differently from a liberal, such as Rosemary Ruether, toward the shared goal of caring for creation. Indeed, the discovery of such a common preference begins with a coincidence, first and foremost, in not coinciding. But this is no surprise. It is a reason for hope. Biblical language is open-ended, caught up in the hermeneutic spiral of sacred story. The play of mythic intentions—the endeavor to close the gap between experience and meaning—assumes the metaphoricity of language. As Bloom puts the point, "To originate anything in language we must resort to a trope, and that trope must defend us against a prior trope. . . . To say and mean something new, we must use language, and must use it figuratively" (1975, 69). Therein lies the power of the metaphor of caring for creation. Metaphor, as anyone who thinks about the matter can appreciate, is essential to the open-endedness of language, to the play of imagination.

In a democratic society such as ours, in which 85 percent of the population professes Judeo-Christian affiliation, the potential political potency of the metaphor of caring for creation is clear. The first language of utilitarian individualism now subtly conditions our private aspirations and dominates our public discourse. Sober-minded observers of both our social ecology (among them Marty, Bellah, and Rorty) and our natural ecology (such as Wilson and Firor) argue that the American experiment teeters on the brink of political and ecological catastrophe. Not all Americans believe our situation to be this dire. Yet who can doubt that the corporate state, the Republic of Special Interests, is a leviathan that threatens to swallow us all? (see MacIntyre 1984, 107). Religious discourse, and especially the second language of the biblical tradition, is our best chance to escape the belly of the whale. Frye argues that politically "we live in subjection to secular powers that may become at any time actively hostile to everything except their own aggressiveness, the leviathan being 'king over all the children of pride' (Job 41:34)" (1981, 190). But the biblical tradition contains within itself the seeds of renewal, the energy to rise up and throw off the monster that envelops us, whether this be Job's eloquent testimony to an irrepressible human dignity and indefatigable sense of responsibility or the parables of Jesus, which subtly yet powerfully undermine the final vocabulary of the state. These stories, as Frye sees so keenly, become exemplary in the context of history "and the specific collision with temporal movement that . . . [biblical] revelation is assumed to make" (198).

Notes

Introduction

1. See Firor 1990 for an authoritative discussion of climate heating and other atmospheric dysfunctions. An exhaustive listing of relevant scientific literature on climate heating, or such other environmental dysfunctions as overpopulation, extinction of species, and deforestation, is beyond the scope of my argument—a few relevant citations must suffice. Full references for these

and other citations throughout the text are found in the Bibliography. Concerning climate heating see Houghton and Woodwell 1989, Woodwell et al. 1983, Kuo et al. 1990, and Post et al. 1990. For arguments against the threat of global warming, see Ausubel 1991. Arguments against climate heating and the imperative nature of timely response are difficult to sustain. Wigley and Raper argue that revised models produced by the Intergovernmental Panel of Climate Change (IPCC), which incorporate such additional variables as feedback from stratospheric ozone depletion and the radiative effects of sulphate aerosols, yield refined and more accurate predictions that anthropogenic climate change is "far beyond the limits of natural variability" (1992, 293). The IPCC predicts an unprecedented rise of 2°F in the next thirty-five years and more than 6° by the end of the next century.

2. See Wilson 1992, Wilson and Peter 1988, Ehrlich and Wilson 1991, Soulé 1991, Erwin 1991, and Meine 1992. See also Kareiva, Kingsolver, and Huey 1992 for essays on interactions between climate change and the extinction of species.

3. Terms like *ecocatastrophe* are well established in English. See, for example, *Random House Webster's College Dictionary* (1991) and *The HarperCollins Dictionary of Environmental Science* (Jones 1992). I employ relevant terminology without further qualification throughout this book.

4. See Hays 1992a. Hays claims that my hypothesis that we should look to religion for guidance in dealing with environmental crisis is not practical. Hays overlooks the importance of democratic debate over the meaning of practice and practicality. His definition of practice assumes that the managing planet Earth approach has been democratically legitimated and that it is the only possible definition of environmentalism. See chap. 1, below, where I argue that by most evidence (for example, increase in atmospheric CO_2, population growth, extinction of species), environmentalism as presently practiced is impractical: clearly, it is not working.

5. The term *sustainability* is used as an alternative to the implicitly oxymoronic term *sustainable development*. See Georgescu-Roegen 1971, Daly 1991, and Ekins et al. 1992 for discussion.

6. See Rorty 1991a for a discussion of skyhooks and toeholds.

7. There are a few exceptions, such as the Church of Jesus Christ of Latter-day Saints. Yet many individual Mormons do care for the Creation, and nothing in principle precludes the church itself from caring for creation.

Chapter 1. Religion in the Context of Ecocrisis

1. For discussions that deconstruct this presumed opposition, see Rorty 1991a, Lyotard 1984, Barbour 1990, MacCormac 1976, and Cobb and Birch 1990.

2. Data on atmospheric CO_2 has been collected at Mauna Loa Observatory since 1958. Ice core samples establish a natural baseline and a variability in carbon dioxide levels (for the past 165,000 years) that are far below the present levels and rate of increase.

3. The controversy over ecological equilibrium continues unabated. See Botkin 1990.

4. See, for example, Perlin 1989.

5. At least in America; the situation may well be otherwise in Europe. Americans yet appear to be a society of faithful people (Caplow et al. 1983). Faith is on the decline in Europe, dramatically so. This is one reason I restrict my study to the United States.

6. See R. Nash 1989, esp. chap. 4, for a historical overview of religion and environmental ethics in North America.

7. I have excluded older texts, such as Bailey's *Holy Earth*, 1916, or Marsh's *Earth as Modified by Human Action*, 1864, since they are not informed by an ecological perspective. But these texts are inconsistent with Livingston's claims, since they recommend a religiously inspired reform of humankind's relation to nature. As Nash (1989) notes, Judeo-Christians were seriously considering their environmental responsibilities from an ecological perspective as early as the 1930s. In 1939, for example, Walter C. Lowdermilk (trained as a forester and hydrologist) addressed the topic of an eleventh commandment—to the effect that thou shalt protect the Creation—over Jerusalem radio. Lowdermilk's eleventh commandment begins "Thou shalt inherit the holy earth as a faithful steward, conserving its resources and productivity from generation to generation" (quoted in Nash 1989, 97).

8. White's essay was originally published in *Science*. Page references herein are to the reprint edition in Barbour 1973.

9. Although White does not mention them in his article, other thinkers reached similar conclusions much earlier. In *Sand County Almanac* (1949), Aldo Leopold explicitly claims that a biblically inspired Abrahamic ethic underlies our misuse of the land. And John Muir took Judeo-Christianity to task before the turn of the century, arguing that narrow-minded religionists could not conceive of the idea that God cared for the rest of creation as well as humankind.

10. See Hargrove 1989 for a critique of Passmore's position.

11. Also see Engel et al., *Ecology, Justice, and Christian Faith*.

12. A consequential question is how to effect behavioral and societal change. Chapter 5, "The Role of the Church," speaks to these issues.

13. And insofar as Wilson's arguments about religion are true, it follows that there is little hope of resolving ecocrisis if either (1) those environmentalists who argue that Judeo-Christianity is the cause of ecocrisis are right, or (2) Judeo-Christians fail to assume their religious obligation to care for creation.

14. Compare Lévi-Strauss 1983 and Evernden 1992b.

15. Despite evolution, but consistent with the first law of matter, the total stock of carbon on earth has not changed, just its distribution. Some ecologists view human overpopulation as a potentially disastrous form of redistributing terrestrial carbon, from many species into one. Increasing human biomass, in other words, is incompatible with biodiversity.

16. For introductory discussions see Turk 1974 and Giddings 1973.

17. See West 1985, esp. 65ff., and Briggs 1992.

18. For discussion see Oelschlaeger 1991, 286–89.

19. Although my subject in this section is the relation of religious discourse to environmentalism generally, and the official or expert discourse of environmentalism specifically, some sociological description is useful. So viewed, environmentalism is diverse, including not only the intellectual communities I discuss but also activists, researchers, and hands-on managers. Popular or grass-roots environmentalists, advocates for a diverse array of conservation philosophies, from radical to conservative, constitute the largest group. Membership in the Sierra Club, for example, reached a high of approximately 800,000 following the Exxon Valdez disaster. The substantive achievements and political influence of such environmental interest groups is considerable and in no way inconsistent with my argument. Religious discourse centering on a metaphor of caring for creation would, if anything, strengthen secular environmentalism. So would additional scientific information. But experts, ensconced in the federal bureaucracy or the modern research university, often distrust grass-roots environmentalists, viewing them as ecologically uninformed tree huggers or radicals who would destroy society to preserve the wilderness. Conservation researchers, including scientists and assisting field workers, constitute another community of interest within the environmental movement. As suggested previously, this group tends to be speechless, occupied primarily with research and efforts to secure continued funding for their projects. Finally, hands-on managers, such as nature interpreters and park personnel, constitute a third identifiable group. As with the environmental research community, this group tends to be silent, in part because of the adverse economic consequences that befall conservation advocates who work for state or federal agencies.

20. Studies relevant to bureaucracy and environmentalism include Weber 1973, Haefele 1973, MacIntyre 1984, Bennett 1987, Johnston 1989, Paehlke and Torgerson 1990, and Bellah et al. 1991.

21. See Leiss 1972 for discussion.

22. The skeptic might reconsider the motivation behind the savings and loan bailout or the federal loans that rescued failed capitalist ventures, such as either Chrysler or Lockheed.

23. The environmental record of the Reagan presidency has already been judged, by conservative and liberal scholars alike, as a failure. Samuel P. Hays, a centrist historian, assesses the Reagan administration as anti-environmental (1987, ch. 15). Notwithstanding its own self-congratulating rhetoric, the Bush administration also failed to achieve a positive record on environmental issues. Even before the Earth Summit in June 1992, the popular press pointed out that Bush's posture as an environmental president was difficult to reconcile with his actions. *Nature*, widely recognized as the most prestigious scientific journal in the world, ran a continuing saga of the environmental misadventures of the Bush administration. The issue for May 21, 1992, captions a lead editorial as "Anti-environment Bush. President George Bush has sided with industry against his own environmental chief on an air pollution issue" (177).

24. Some of my readers suggest that a global ecological disaster might help transform American culture. The problem is that we already live among the ruins of ecological disaster. How much destruction is required before people will respond?

25. John F. Kennedy was subject to suspicion because of his Roman Catholicism. His campaign broke through in West Virginia, where he publicly avowed that if elected, he would fulfill his responsibilities as president even if they conflicted with the dictates of his faith. More recently, Jimmy Carter was suspect because of his "fundamentalism." The Reagan and Bush presidencies seem to be an exception, due to the rise of the Moral Majority. But the point is firm: neither Reagan nor Bush was perceived as a religious candidate per se, as was, for example, Pat Robertson.

26. The Vatican reaffirmed its opposition to any method of birth control other than abstinence immediately before the Earth Summit Conference in June 1992. For thoughtful criticism of the Vatican's position see Schwarz 1993.

Chapter 2. Religion and the Politics of Environmentalism

1. Some theorists, such as Lance deHaven-Smith (1991), are less sanguine about polls that indicate public support for environmentally oriented policy. Arguing as a theorist of public opinion, deHaven-Smith contends that there is no commitment among the mass public to an environmental movement (5).

2. See Milbrath 1989, 118–24, for a discussion and statistical analysis of competing paradigms of belief (the so-called new environmental paradigm and the dominant social paradigm). I am indebted to Milbrath and to Miller 1991, 26–28, for suggesting the utility of tabular display for purposes of comparison-contrast and for relevant overviews of the differences between the DSM and the NSM. See also Paehlke 1989, 144–45, and Naess 1989, 16, for useful summaries.

3. Self-interest does not precisely equate with private interest, since individuals can have a self-interest in the common good, that is, individuals can have an other-interest. Private interest is just that, private. Human beings are biologically encoded and culturally empowered to act in their own self-interest (although we are sometimes unclear as to what is in our self-interest). Private interest, as I use the term, is the attempt to convert one kind of self-interest, say in short-term economic benefits as measured by monetary income, into the "public interest." For example, as a farmer raising wheat in the semiarid corner of southwestern Kansas, my private interest lies in getting the government to drill wells and provide energy to pump groundwater to grow wheat that becomes a glut on the market—all justified in the name of the public interest. Powerful special interests, such as those representing agribusiness, exert their influence through massive publicity and media blitzes and especially through political donations. But no viable conception of the common good can be built around the idea that it is an aggregate of private interests: in any complex system, the whole is always more than the sum of its parts. In an ideal world or "best of all possible worlds," a narrow calculus of value that what is good for me might fortuitously coincide with what is good for you and everyone else. In the real world, a calculus of value restricted to what is good for me (privately) is often not good for you. I might profit by dumping deadly chemicals on the road you drive home on. Or by getting the government to subsidize or otherwise finance my business.

4. See Vig and Craft 1990.

5. There are many difficulties with cost-benefit analysis beyond the scope of the present discussion, ranging from actual errors of measurement to difficulties in trying to find adequate criteria of measurement.

6. In a United Nations survey conducted by Louis Harris and Associates in 1989, 75 percent of the world's people were worried about environmental degradation and believed that more stringent measures to protect the environment were appropriate.

7. A number of interesting questions involved in this position go beyond the scope of my argument. For example, how do individuals come to hold public values? Do individuals have selfish interests in the common good? Are values that everyone holds in common by that fact alone not private preferences? Or are private preferences, held by a majority of voters, also called public values? Some readers suggest that linking self-interest with other-interest is a more effective strategy than arguing for public values, on the grounds that any individual interest in intergenerational equity or in the preservation of endangered species remains a self-interest, albeit an altruistic one. Further, it occurs to these same readers that I can have a self-interest in the common good that goes beyond the market, that is, any selfishly economic or anthropocentric calculus of value. Sagoff's terminology, which distinguishes "citizen preferences" from "consumer preferences," largely avoids this issue. Dawkins (1976) argues that self-interest does not preclude other interest; indeed, it may have a survival advantage, ensuring perpetuation of "the selfish gene."

8. See Wright 1992 for an excellent discussion.

9. I am not advocating, as some skeptics might assume, either a return to theocracy or the construction of the New Jerusalem or its secular equivalent.

10. Adam Smith, ironically, argues in his *Theory of Moral Sentiments* (1759) that humans are more than greedy little pigs, since they have "fellow feelings" of sympathy toward others that balance their self-interest. But he gives the market entirely over to economic self-interest. In Smith's opinion the invisible hand, and not fellow feelings, sees to the common good. Smith's work still has enormous influence, but it has many, many problems, not the least of which is the existence of the "invisible foot." That is, not all the unplanned outcomes of individuals seeking their own economic self-interest are socially beneficial; many are ecologically and intergenerationally disastrous. Collective choices or citizen preferences, to use Sagoff's words, cannot be reduced to consumer preferences.

11. See the many critiques of the objectivity of economics, especially Georgescu-Roegen 1971, McCloskey 1985, Daly 1991, and Ekins et al. 1992.

12. On this interpretation of Smith see Heilbroner 1972 and Oelschlaeger 1991, esp. 91–94.

13. Paehlke's position is neither that economic growth nor technological progress is intrinsically bad. He would favor, for example, the growth of a solar-powered economy and the associated technology. The point is that some economic growth, such as that fueled by hydrocarbon-based energy or the destruction of the Amazon rainforest, is a recipe for ecological disaster.

14. Corporations are not, however, always profit maximizers. Managers often value their perquisites over corporate interests, as evidenced by the enormous salaries paid to chief executive officers of firms losing millions or even hundreds of millions of dollars annually.

15. The situation is not the same in such democracies as Canada, where campaign funding is closely regulated so that private interests do not achieve undue influence.

16. See the excellent discussion in Bellah et al. 1991, 287–306. Humans do form institutions, but institutions also shape human beings individually and culture more generally.

17. Almost 35 million Americans lack adequate health care.

Chapter 3. The Sacred Canopy

1. The notion of cultural overdetermination does not entail the conclusion that genes are irrelevant. The point is that our socially learned behavior makes us distinctively human, both culturally and individually. Every human being participates uniquely in the human gene pool, although it is true to say that the human gene pool is universally shared among the members of our species. But cultural diversity illustrates how distinctive patterns, such as religious behavior, are permitted regardless of genetic makeup. It is perhaps more accurate to emphasize (along with Geertz) the idea that genes underdetermine human behavior than to say that they permit cultural variability. In any case, extrinsic sources of information (beyond the genetic code) are crucial to determining human beingness. Within a culture, learned behavior introduces behavioral variability both across groups and across persons within groups.

2. Some interesting questions arise concerning differences between sociohistorical and biogenetic (Darwinian) evolutionary processes. There are also, according to Lumsden and Wilson (1981, 1983) interactions among memes and genes, that is, coevolutionary processes. Such issues, however important, are largely beyond the scope of my inquiry. See chap. 6, n. 7.

3. Some readers, especially those with conservative commitments, asked for the incorporation of relevant biblical passages such as this. Chapters 4 and 6 provide amplifying details tied to biblical passages. However, detailed biblical study and exegesis apropos of my theme that the Bible provides the woof and warp of Western civilization are outside the focus of my argument.

4. For example, see Bock 1980.

5. Rue's thesis is supported by cognitive science and psychology as well as some theories of artificial intelligence. For introductory discussions see Schank 1990 and Fischler and Firschein 1987. Schank argues that human beings define themselves through story. Further, according to Schank, cognition—or at least the primary features of thinking such as data collection and manipulation, comprehension, explanation, planning, communication, and integration—is dependent on story. Intelligence, he explains, "for machines as well as for humans, is the telling of the right story at the right time in the right way" (241–42).

6. See Bloom 1990 for amplifying discussion.

7. In the United Kingdom, Prance is probably better known than Wilson. He has a substantive research record, an extensive list of research publications, and is recognized as a major contributor to rainforest preservation, especially through his theory of phytogeography (see Foresta 1991, 37ff.). More recently Prance's work in ethnobotany has figured prominently in efforts to conserve Amazonia. See Prance 1990a, 1990b.

8. See Evernden 1985, 1992b.

9. See among others Tawny [1926] 1947, Troeltsch [1912] 1987, and Weber [1904–5] 1958.

10. See Ekins et al. 1992, 64ff.

11. One recalls the popular bumper sticker of the 1970s proclaiming, "If it feels good, do it!"

12. Ortega's (1932) distinction between the mass person and the select person is useful, because the select person recognizes the obligation to uphold the institutions of liberal democracy through service. The mass person, in contrast, is the "spoiled child of history" who, as is characteristic of adolescents, thinks the world owes him or her a living. Such an individual believes above all else in the life of comfort and convenience.

13. See Oelschlaeger 1991, esp. chaps. 1 and 2, Glacken 1967, and Johnston 1989.

14. Some scholars, such as Paul Shepard (1982), believe this fact alone fundamentally undercuts the viability of the Judeo-Christian narrative tradition. Though I agree with much of Shepard's analysis, I believe he underestimates the interpretive possibilities inherent in the Great Code.

15. See Galbraith 1952 for discussion.

16. The three members of the Council of Economic Advisors are perhaps the most influential of this elite cadre. The New York Times News Service reports (*Dallas Morning News*, January 4, 1993, section D) that President Bill Clinton appointed Alan Blinder to the Council of Economic Advisers after criticism from mainstream economists that the chair of the council, Laura Tyson, lacks "the necessary analytical skills" to assess economic policy. Macroeconomists at "elite" eastern institutions, among them Harvard, MIT, and Princeton, "view the council as their chief means of influencing administration policy." They also view the council as their Washington "embassy."

17. Macroeconomics is a mathematical discipline that employs computer models to simulate the quantifiable monetary effects of economic policy. The problem with economic analysis generally, and cost-benefit analysis specifically, is that "there is a contentious subjective element in regard to who decides which costs and benefits are to be considered, and whether or not they represent particular interest groups" (Jones 1992, 91).

18. See Kristeva [1981] 1989, 38–39, for a brief account of metaphor and meaning. She argues there, as I do here, that semantics and rhetoric overlap.

19. Nietzsche writes this: "What then is truth? A movable host of metaphors, metonymies, and anthropomorphisms: in short, a sum of human relations which have been poetically and rhetorically intensified, transferred, and embellished, and which, after long usage, seem to a people to be fixed, canonical, and binding. Truths are illusions which we have forgotten are illusions; they are metaphors that have become worn out and have been drained of sensuous force, coins which have lost their embossing and are now considered as metal and no longer as coins" (1872–73, 891).

20. I am indebted to Susan Bratton for this line of analysis. In a forthcoming study she provides a number of illustrations showing that biblical models, such as the good neighbor (there are many relevant biblical passages, such as Lev. 19.18), lead people to care for creation.

21. This is not incomprehensible. See McCloskey 1985.

22. Alternatives to top-down forms of government and large-scale bureaucratic organizations are discussed by Berg 1978, Dryzek 1987, Snyder 1990, and Andruss et al. 1990. My inquiry focuses on politically empowering a social movement toward sustainability rather than on the specific forms of sociopolitical organization a sustainable society might take.

23. See Clark 1989.

24. See Steidlmeier 1993.

25. *Science* published an essay claiming that the United States would collapse before 1980.

26. See Ekins et al. 1992, Costanza 1992, MacNeil et al. 1992, Krabbe and Heijman 1992, and Meadows et al. 1992. Also see the *Journal of Ecological Economics.*

Chapter 4. The Spectrum of Belief

1. I emphasize the words "might serve." An anonymous referee argues in commenting on a draft of this book that all traditions of faith do not present equal possibilities to care for creation. This reader points to specific denominations, like the Church of Jesus Christ of Latter-day Saints, that have denied that environmental issues are part of their mission, to fringe sects like the one established by James Jones, and to individuals like James Watt (who justified his anti-environmentalism in the name of dominion over the earth) as challenging my thesis. My argument, however, does not require ecotopian uniformity but only the possibility of solidarity. Further, James Jones was insane; my argument is inclusive to the extent that it recognizes traditions of faith that are not self-defeating. Finally, people of faith are typically more aware of the insufficiencies of their beliefs then those who stand outside as nonbelievers. I see no point in criticizing religion for its failures, since that does not bear on my thesis. But I can criticize various faith traditions for bad faith, for the failure to care for creation as one's faith requires. At the conservative end of the religious spectrum (see figure 1), bad faith equates with sin and at the liberal end it equates with inauthenticity. See Bratton 1983 for a discussion of bad faith in the case of James Watt, secretary of the Interior during the Reagan administration, a religious conservative who in claiming the right to exploit nature demonstrated his theological illiteracy.

2. Political leverage is differentially distributed. Religious conservatives, acting through such organizations as the Moral Majority, have achieved considerable influence in recent elections. Liberals, too, with a tradition of activism (God goes to Washington) have achieved considerable success. Groups with small numbers of affiliates, such as Jews (less than 2 percent of the population), sometimes have considerable influence due to political activism and sophistication. See Massey 1991 and J. Nash 1991 for overviews of Christians and environmental politics.

3. There are, as mentioned, exceptions, since some conservatives translate the Hebrew *yom* (day) as a "period of time." This translation avoids difficulties inherent in the literal meaning of the term *day*.

4. Some biologists raise the question of cosmic teleology but find no evidence for it. See Dawkins 1987.

5. Others, such as George Hendry (1980), set out a different strategy that aims to recover ground lost from science. Hendry argues that the concept of nature is a scientific construct that makes "the world the property of man and not of God" (197). But the concept of nature cannot substitute for a concept of creation, if for no other reason than human beings have no real home in the objectively defined world of scientific nature. Science is power, Hendry admits, "But the pragmatic triumph of science has been purchased at the price of an alienation of man from nature" (197). What solution does Hendry envision? Primarily a recovery of some biblical sense of the human place in nature. "When Paul speaks of Christians as having the Spirit, he is referring to the knowledge they have through the Spirit that their lives are involved in the process of death and resurrection, which has its paradigm in Christ, and which is being reproduced in the world of nature" (214). This opens the door to a sympathy with nature, now known as the entirety of creation rather than as a mere resource for human appropriation. "Sympathy with the sufferings of living creatures, whether they be human beings or such animals as human beings are capable of forming some kind of relationship with, is accepted as something incumbent upon Christians, who owe their being as Christians to the suffering of Christ *for* them" (215).

6. Metaphysically considered, it can be argued that science and religion overlap. For example, see S. Alexander 1979.

7. It is also difficult to imagine the Scientific Revolution occurring apart from the influence of Presocratic and Greek thought. The Greek word *kosmos* means order or world.

8. Creation stories are legitimating narratives that enable a *sacred canopy*, the overarching web of beliefs that provide ultimate justification for life, that offer a reason to be, and that undergird a sense of human dignity. Of course, some of these narratives—such as conservative ones that view "man" as created in the image of God—preclude thinking of ourselves as storytelling culture-dwellers. And yet the actual behavior of conservative Christian communities shows how fundamentally important their stories are, since they literally learn how to be Christians by telling and retelling biblical stories. Philosophical and theological argumentation is secondary within these faith traditions.

9. See Lawson 1985 for an able introduction.

10. Italics indicate usages consistent with the linguistic conventions of religious conservatives.

11. Rosemary Ruether, for example, argues that the traditional Judeo-Christian narratives marginalize women and nature, leading to the subjugation and exploitation of both.

12. See Victor 1991.

13. Compare the Good News Bible translation: "The world and all that is in it belong to the Lord; the earth and all who live on it are his."

14. This position is consistent with contemporary ecophilosophy, where much of the professional debate turns on intrinsic value. Schaeffer, however, derives his argument from a biblical rather than a philosophical narrative tradition. His position also resonates with that of moderate Protestants, such as David Griffin, who also recognize intrinsic value and a natural hierarchy but, unlike Schaeffer, accept the reality of evolution—that is, the legitimacy and importance of science-based story sources.

15. Although it is beyond the scope of my argument, there is an enormous literature on this theme. Among many, see Heidegger 1977b, Ihde 1979, Marx 1964, and Mander 1991.

16. Calvin DeWitt holds a Ph.D. in biology from the University of Michigan and is a professor of environmental studies at the University of Wisconsin–Madison. He is also director of the Au Sable Institute of Environmental Studies in Mancelona, Michigan.

17. Public lecture, Feb. 27, 1992, Denton, Texas.

18. See Jakowska 1986, 132ff., for further discussion.

19. Critics contend that in the Third World the sheer pressure of population rather than consumerism destroys natural ecosystems. They also criticize the Vatican for refusing to approve any method for birth control other than periodic abstinence. For an examination by Catholics of Vatican policies on global population see the March 6, 1993, issue of *America*. Critics outside Catholicism should note that the Holy See "admits the reasonableness of formulating an official policy on population growth in a nation" and "lends official sanction to such efforts as the family-planning clinics sponsored in a growing number of dioceses in the United States." See Paul 1965, 302, nn. 270, 271.

20. This terminology can create confusion, since *Conservative Judaism* exists as an actual North American movement, in addition to *Reform Judaism*, categorized here as liberal, and *Orthodox Judaism*, categorized as conservative. In terms of figure 1, *Conservative Judaism* is moderate. It should also be noted that within each grouping there is considerable diversity; for example, no one national institution is recognized by all Orthodox Jews.

21. Equally offensive is the lack of funding to enable full implementation of the Endangered Species Act. Approximately three thousand species await endangered species status, more than a thousand of them held up in one agency alone (U.S. Fish and Wildlife Service). Wise-use conservationists also like to argue that the Endangered Species Act threatens the economy. In truth, less than 0.1 percent of all development projects studies by the U.S. Fish and Wildlife Service (23 of 34,600 programs) were found to have adverse impacts.

22. Ehrlich (1977) is a well-known scientific advocate of population control; also see Ehrlich and Ehrlich 1990. Hardin is known for his controversial book *Exploring New Ethics for Survival* (1972).

23. Both religionists and scientists, embedded in their own traditional stories, tend to reject Teilhard's message categorically. This rejection is comprehensible, since his narrative is an alternative to either a conservative (or even moderate) creation story or to a strictly scientific account of evolution.

24. McDaniel's paper "Emerging Options in Ecological Christianity: The New Story, the Biblical Story, and Panentheism" (1991) makes this difference evident.

25. See Cobb and Tracy, *Talking about God* (1983).

26. Berry's work has been an inspiration to many, including Jay McDaniel, Michael Dowd, and Brian Swimme. Dowd (1991), a pastor for the United Church of Christ, adapts Berry's work for Protestants in *Earthspirit*. Brian Swimme, a physicist also involved in popularizing the new cosmology, recognizes Berry as the first to realize "the cultural and planetary significance of a common creation story" (1988, 56, n. 7).

27. Prigogine's *Order Out of Chaos* (1984) is an excellent introduction to the new science. Also see Popper and Eckles 1977 and Mayr 1988, esp. 38–66. The utility of the new science is that teleonomic (or some forms of teleological) explanation is scientifically legitimate, since no appeal to supernatural agency (or a principle of creation external to the created) is required.

28. Lovelock's arguments (1989, 1990) in support of the Gaia hypothesis also lend credence to Goddess feminists. He argues that the earth cannot be reduced (in the prevailing fashion of science) to simply matter-in-motion.

29. A standard introductory text is Spencer et al. 1977.

Chapter 5. The Role of the Church

1. See Boyte 1984, 1986a, 1986b; Barber 1984, 1986; and Lappé 1990 for discussion.

2. MacIntyre (1984) also envisions a way around this problem (258ff.) that is consistent with my argument—namely, the rhetorical tradition offers a plausible solution to the failure of ethical philosophical discourse, since conversation is the primary means by which a community can work toward a good society.

3. Hargrove's (1989) arguments in some ways parallel MacIntyre's, though his derivation is different. Hargrove argues that "the rise of emotivism out of logical positivism" has tended to undercut environmental ethics, not only because of the diversity in environmental ethics but because economic judgments of environmental value have, in comparison to those of environmental ethics, appeared to be objective (210).

4. The Marxian critique of religion suggests that, even into the nineteenth century, religion was an opiate that quelled opposition to the state. Yet Marxians specifically, and political theology more generally, have tended to ignore the minority traditions (for example, Gnosticism) within Judeo-Christian culture that were always a thorn in the side of secular power.

5. In economic theory this situation is reflected by the exclusion of nonmonetary income, such as the services of individuals in the household (largely women), public interest groups, and the biosystem from the National Income Accounts. As Ekins et al. point out, "It is arguable that such groups [and biosystem services] do as much for our quality of life as the formal business sector" (1992, 68). The consequence is that very narrow and exclusively monetary measures of wealth are presented to the nation as an index of social welfare.

6. The issue is not the market per se but the attempt to reduce environmental issues to consumer preferences and the question "How much are you willing to pay?" Citizen preferences in, for example, biodiversity, go beyond consumer preferences for inexpensive hamburgers and plywood derived from the short-term exploitation of South American rainforests. See Sagoff 1988 for discussion.

7. The family is an institution included within the present study only to the extent that it is connected with the church. There is no doubt that the family, our most intimate institution, powerfully influences human character.

8. Adapted from Elmer-Dewitt 1992.

9. The Forest Service leads the assault on America's ancient forests and ecosystems. Masquerading under a "multiuse" policy, the Forest Service functions as an administrator for rich and powerful private corporations, expending public funds to subsidize the harvest of timber—primarily through the construction of roads (that have devastating ecological implications)—that otherwise would be priced out of the market. For example, prime logs, cut from Alaskan old-growth forests, are shipped to Japan by for-profit timber companies carrying a subsidy from the American taxpayer of $1.50 each. The Department of Agriculture, operating through the so-called Animal Damage Control Agency (ADCA), spends tens of millions of dollars each year to exterminate wildlife, including black bears, mountain lions, coyotes, gray and arctic foxes, egrets, herons, owls, and snapping turtles. Ostensibly in the public interest, the ADCA is a welfare system for agribusiness. So, too, the Army Corps of Engineers has served as a welfare program for developers, using public tax dollars to build the water resource projects that provide the infrastructure that supports growth.

10. The Pentagon has bombed, mined, and shelled millions of acres, primarily in the American Southwest, but elsewhere as well, into ecological oblivion. The legend of the misadventures of the Rocky Flats Nuclear Weapons Plant continues to grow in Colorado; the illegal dumping and disposal of radioactive wastes pose as yet unresolved issues (and the Justice Department is being investigated to see if its own prosecutors enforced the law or "cut a deal" with Rockwell International, owner of the Rocky Flats facility). The EPA estimates Rocky Flat cleanup costs at $1 billion. More generally, the Government Accounting Office estimates the cleanup costs for damage to the environment by the Department of Energy at $90–120 billion. In the Pacific Northwest, the Atomic Energy Commission intentionally released a cloud of radioactive iodine five hundred times as lethal as that which escaped from the more recent Three Mile Island power plant disaster. No advance warning was provided to the public affected by the release—literally an experiment carried out by the American government on thousands of unwitting citizens. Or consider that the EPA itself has been convicted of dumping toxic wastes. And the story continues ad nauseam at the National Park Service and Bureau of Land Management.

11. Michael Nieswiadomy (1992), a resource economist, argues that U.S. Forest Service policies are not only an ecological but an economic disaster. Conservative political commentators have taken the conservative movement (and the New Right) to task for not addressing pressing ecological problems (Fleming 1990). Similarly, Donald Worster (1985), a liberal environmental historian, details how the Bureau of Land Management, with the blessing of Congress, manages vast western acreage for the benefit of privately owned cattle and sheep empires—essentially a welfare program that provides some benefits for the middle class while the larger share goes to the wealthy and powerful.

12. More recently, the Bush administration provoked conservation groups, such as the Sierra Club and the Wilderness Society, to become outspoken critics of its environmental record. The Bush shift on wetlands, a semantically based rather than scientifically grounded redefinition that opened up protected areas to economic development, turned into a public relations disaster. Similarly, the decision to cut old-growth forest that provided habitat for the endangered northern spotted owl severely damaged the administration's environmental credibility. And the president's misrepresentation of himself as "the environmental president" was particularly upsetting to environmental groups, who viewed such statements as propaganda. *Newsweek* characterized Bush as "The Grinch of Rio" for his actions (or lack of) at the United Nations Conference on the environment held at Rio de Janeiro in 1992. The Bush administration refused to sign the treaty on biodiversity and signed the treaty on CO_2 emissions only after eliminating clauses that called for specific amounts and kinds of greenhouse gas reductions. The other industrialized democracies, including the European nations and Japan, signed a separate treaty mandating the stabilization of greenhouse emissions at 1990 levels by the year 2000 (Begley et al. 1992, 30).

Members of the Bush administration have also been criticized by environmentalists and public interest groups. Vice President Quayle's Council on Competitiveness, empowered to overturn policies that jeopardize short-term profit and "economic progress," was rebuked in 1992 for its violation of a basic precept of Western law: namely, there are no exceptions. The Council on Competitiveness granted corporations the right to violate air quality laws without either sanction or the responsibility to notify the public. Secretary of the Interior Manuel Lujan enraged environmentalists by arguing that the jobs of loggers were more important than the life of an endangered species, the spotted owl. Lujan even suggested that the Endangered Species Act should be repealed. And John Sununu, Bush's chief of staff until mid-1991, was lampooned by the scientific press for his now infamous pronouncement that "methane is not a greenhouse gas." More generally, during the past decade environmental laws have repeatedly gone unenforced because of underfunding of the agencies empowered to enforce them. And officials who have attempted to enforce the law have been harassed, moved to new locations, demoted, and in some cases fired.

13. Actually, styrofoam can be recycled. McDonald's officials knew this, but public opinion was so overwhelmingly negative on styrofoam that the switch to paper was deemed the politically correct action. One advantage of paper over styrofoam is its biodegradability—but paper buried in landfills resists biodegradation.

14. Universities figuratively represent the entire educational complex, from primary schools up. There is some evidence that primary schools are helping children develop the skills to comprehend ecological problems. (On the other hand, our educational system also has its critics.) My discussion is limited to the university. Unlike the university, grade schools and high schools do not claim to offer expertise (experts, knowledge, technology) for resolving ecocrisis.

15. The vocabularies of deep ecology, ecofeminism, and bioregionalism are other alternatives. But they are unfamiliar languages to most Americans, while the Great Code is a lingua franca at least in its broad outlines, such as the creation story, the covenant tradition, the story of Joseph and Mary, the Lord's prayer, and the story of Jesus of Nazareth.

16. Concerning the economic effects of environmentalism, see Meyer 1992, 1993, and Meyer, *Environmentalism and Economic Prosperity* (provisional title, forthcoming from MIT Press). Meyer argues that environmental policy has in the aggregate no negative economic consequences. "If environmentalism does have negative economic effects they are so marginal and transient that they are completely lost in the noise of much more powerful domestic and internal economic influences" (1993, 10).

17. Reprinted in Birch et al. 1990.

18. See Granberg-Michaelson 1992 for a report on the WCC and the Rio Earth Summit. I am indebted to an anonymous referee for this reference and other useful information in this section.

19. See Roberts and Amidon 1991 for an ecumenical selection of earth prayers.

20. See Beeman 1989.

21. The terms culture of coherence and culture of separation are used by Bellah 1991. Clearly a culture of coherence would address issues of equity, civil rights, health care, universal education, and social justice.

22. Local churches typically find their places within denominations, higher levels of organization that provide various kinds of support for the local church in its activities, whether this is organizing missionary work, running seminaries, supporting parachurch organizations in the national capital, or operating a publishing house.

Chapter 6. Redescribing Religious Narrative

1. Among many see Manfred Max-Neef et al. 1990.

2. The politics of sustainability does not require a literal political majority, that is, a majoritarian green party. A committed minority of green voters, perhaps no more than 20–30 percent voting as a bloc, would empower either a Republican or Democratic president to take the political lead on sustainability. No presidential candidate could ignore such a formidable voting bloc. Consider the implications of the 1992 elections. According to exit polls (November 3, 1992) conducted by ABC News, only 6 percent of the voters cared about environmental issues (compared to 43 percent who ranked the economy as a priority issue); but among the green voters, 73 percent went to the Clinton-Gore ticket (compared to 52 percent of the "economic issues" voters) (Fineman 1992, 10). The Clinton-Gore margin of victory was less than five points; clearly, green voters were indispensable. Election of a "green Congress" involves a similar kind of logic. Perhaps twenty senators and eighty representatives are required to empower movement toward sustainability; if the faithful begin to care for creation, then green senators and representatives might even occasionally be elected from states whose economies are tied to resource extraction, since these voters would countervail anti-environmental voters.

3. Both philosophers of religion and theologians can agree on this. Compare, for example, Collingwood 1940 and Schaeffer 1968.

4. See Leiss 1972 for discussion.

5. There are other possibilities, including Jungian, Bergsonian, Adlerian, and Jamesian approaches. However, the treatment of Freud by both Bloom and Rorty makes Freudianism useful in the present context.

6. See Schneidau 1976.

7. Rorty's pragmatism and mine are similar but not the same. I believe that although we belong to language, we also condition its development. Rorty believes that language evolves in a Darwinian fashion; in my opinion, he goes too far with his Darwinian metaphors. Human beings inhabit stories (an evolutionary niche) that evolve in ways conditioned by learned behavior, that is, the history of effects. As Dawkins and Wilson suggest, culture is an evolving structure of memes (texts or textual traditions) rather than chance mutation. A culture's story, of course, remains subject to "natural selection," as environmental crisis and the threat of ecocatastrophe imply.

8. An anonymous referee suggested this reading to me. Roderick Nash also argues for this interpretation (1989, 92).

9. The play of language (imagination) also expresses itself through prophets and poets, litterateurs and philosophers, scientists and mythographers.

References

Adam, David. 1987. *The Cry of the Deer: Meditations on the Hymn of St. Patrick*. Wilton, Conn.: Morehouse-Barlow.

Albanese, Catherine L. 1990. *Nature Religion in America: From the Algonkian Indians to the New Age*. Chicago: University of Chicago Press.

Alexander, Charles P. 1992. "Gunning for the Greens." *Time* (February 3), 50–52.

Alexander, Samuel. [1920] 1979. *Space, Time, and Deity*. Gloucester, Mass.: Peter Smith.

Andruss, Van, Christopher Plant, Judith Plant, and Eleanor Wright, eds. 1990. *Home! A Bioregional Reader*. Philadelphia: New Society.

Aquinas, St. Thomas. 1952. *Summa Theologica*, trans. Fathers of the English Dominican Province, rev. Daniel J. Sullivan. In Robert Maynard Hutchins, ed., *Great Books of the Western World*, vol. 19. Chicago: Encyclopedia Britannica.

Argyle, Michael. 1987. *The Psychology of Happiness*. London: Methuen.

Aristotle. 1941. *The Basic Works of Aristotle*, ed. Richard McKeon. New York: Random House.

ASCEND. 1991. "Conference Statement: International Conference on an Agenda of Science for Environment and Development into the Twenty-first Century (ASCEND 21)." Typescript, Vienna.

Austin, Richard Cartwright. 1988. *Beauty of the Lord: Awakening the Senses*. Atlanta, Ga.: John Knox Press.

Ausubel, Jesse H. 1991. "A Second Look at the Impacts of Climate Change." *American Scientist* 79:210–21.

Backman, Mark. 1987. "Introduction: Richard McKeon and the Renaissance of Rhetoric." In McKeon, *Rhetoric*.

Barber, Benjamin R. 1984. *Strong Democracy: Participatory Politics for a New Age*. Berkeley: University of California Press.

———. 1986. *The Conquest of Politics: Liberal Philosophy in Democratic Times*. Princeton: Princeton University Press.

Barbour, Ian G. 1973. *Western Man and Environmental Ethics: Attitudes Toward Nature and Technology*. Reading, Mass.: Addison-Wesley.

———. 1990. *Religion in an Age of Science: The Gifford Lectures, 1989–1991*, vol. 1. San Francisco: Harper & Row.

Barlow, R. M. 1989. "Buber Looks at Nature: An Alternative Epistemology." *Contemporary Philosophy* 12, 10:5–11.

Beeman, Larry F. 1989. "Caring for Creation." *Teacher in the Church Today* (October), 4–7.

Begley, Sharon, Brook Larmer, Mac Margolis, and Daniel Glick. 1992. "The Grinch of Rio." *Newsweek* (June 15): 30–32.

Bellah, Robert N., Richard Madsen, William M. Sullivan, Ann Swidler, and Steven M. Tipton. 1985. *Habits of the Heart: Individualism and Commitment in American Life*. New York: Harper & Row.

———. 1991. *The Good Society*. New York: Alfred A. Knopf.

Bemporad, Jack, ed. 1977. *A Rational Faith: Essays in Honor of Levi A. Olan*. New York: KTAV.

Bennett, James. 1987. *Unthinking Faith and Enlightenment*. New York: New York University Press.

Berg, Peter. 1978. *Reinhabiting a Separate Country: A Bioregional Anthology of Northern California*. San Francisco: Planet Drum Foundation.

Berger, Peter L. 1967. *The Sacred Canopy: Elements of a Sociological Theory of Religion*. Garden City, N.Y.: Doubleday.

Berger, Peter, and Thomas Luckmann. 1966. *The Social Construction of Reality: A Treatise in the Sociology of Knowledge*. Garden City, N.Y.: Doubleday.

Bernards, Neal, ed. 1991. *The Environmental Crisis: Opposing Viewpoints*. San Diego: Greenhaven.

Berry, Thomas. 1987a. "The Earth: A New Context for Religious Unity." In Lonergan and Richards, eds., *Thomas Berry and the New Cosmology*.

———. 1987b. "Economics: Its Effects on the Life Systems of the World." In Lonergan and Richards, eds., *Thomas Berry and the New Cosmology*.

———. 1987c. "Our Future on Earth: Where Do We Go from Here?" In Lonergan and Richards, eds., *Thomas Berry and the New Cosmology*.

———. 1987d. "Twelve Principles: For Understanding the Universe and the Role of the Human in the Universe Process." In Lonergan and Richards, eds., *Thomas Berry and the New Cosmology*.

———. 1988. *The Dream of the Earth*. San Francisco: Sierra Club Books.

———. 1991. *Befriending the Earth: A Theology of Reconciliation Between Humans and the Earth*, with Thomas Clark, Stephen Dunn, and Anne Lonergan, eds. Mystic, Conn.: Twenty-third Publications.

Berry, Thomas, and Brian Swimme. 1992. *The Universe Story: From the Primordial Flaring Forth to the Ecozoic Era—A Celebration of the Unfolding of the Cosmos*. San Francisco: HarperCollins.

Birch, Charles, and John B. Cobb, Jr. 1981. *The Liberation of Life: From the Cell to the Community*. Cambridge: Cambridge University Press. Reissued by Environmental Ethics Books, Denton, Tex., 1990.

Birch, Charles, William Eakin, and Jay B. McDaniel, eds. 1990. *Liberating Life: Contemporary Approaches to Ecological Theology*. Maryknoll, N.Y.: Orbis Books.

Bizzell, Patricia, and Bruce Herzberg, eds. 1990. *The Rhetorical Tradition: Readings from Classical Times to the Present*. Boston: Bedford Books of St. Martin's Press.

Blackstone, William T., ed. 1974. *Philosophy and Environmental Crisis*. Athens: University of Georgia Press.

Bloom, Harold. 1975. *A Map of Misreading*. New York: Oxford University Press.

———. 1989. *Ruin the Sacred Truths: Poetry and Belief from the Bible to the Present*. Cambridge: Harvard University Press.

———. 1990. *The Book of J*, trans. David Rosenberg. New York: Grove Weidenfeld.

Bock, Kenneth. 1980. *Human Nature and History: A Response to Sociobiology*. New York: Columbia University Press.

Bohm, David. 1957. *Causality and Chance in Modern Physics*. Philadelphia: University of Pennsylvania Press.

Bookchin, Murray. 1990. *Remaking Society: Pathways to a Green Future*. Boston: South End Press.

Botkin, Daniel B. 1990. *Discordant Harmonies: A New Ecology for the Twenty-first Century*. New York: Oxford University Press.

Bowman, Douglas C. 1990. *Beyond the Modern Mind: The Spiritual and Ethical Challenge of the Environmental Crisis.* New York: Pilgrim Press.

Boyte, Harry C. 1984. *Community Is Possible: Repairing America's Roots.* New York: Harper & Row.

Boyte, Harry C., and Frank Riessman, eds. 1986a. *The New Populism: The Politics of Empowerment.* Philadelphia: Temple University Press.

Boyte, Harry C., Heather Booth, and Steve Max. 1986b. *Citizen Action and the New American Populism.* Philadelphia: Temple University Press.

Bramwell, Anna. 1989. *Ecology in the Twentieth Century: A History.* New Haven and London: Yale University Press.

Bratton, Susan Power. 1983. "The Ecotheology of James Watt." *Environmental Ethics* 5:225–36.

———. 1988. "The Original Desert Solitaire: Early Christian Monasticism and Wilderness." *Environmental Ethics* 10:31–55.

———. 1992a. "The 'New' Christian Ecology." In Oelschlaeger, ed., *After Earth Day.*

———. 1992b. *Six Billion and More: Human Population Regulation and Christian Ethics.* Louisville, Ky.: Westminster/John Knox Press.

———. 1993. *Christianity, Wilderness, and Wildlife: The Original Desert Solitaire.* Scranton, Penn.: University of Scranton Press.

Briggs, John. 1992. *Fractals: The Patterns of Chaos.* New York: Touchstone.

Brockway, George. 1985. *Economics: What Went Wrong, and Why, and Some Things To Do About It.* New York: Harper & Row.

Brown, Lester R., et al. 1989. *State of the World 1989: A Worldwatch Institute Report on Progress Toward a Sustainable Society.* New York: W. W. Norton.

———. *State of the World 1990: A Worldwatch Institute Report on Progress Toward a Sustainable Society.* New York: W. W. Norton.

Buber, Martin. [1923] 1970. *I and Thou,* trans. Walter Kaufmann. New York: Charles Scribner's Sons.

Bunge, Robert. 1984. *An American Urphilosophie: An American Philosophy BP (Before Pragmatism).* Lanham, Md.: University Press of America.

———. 1988. "Community: Key to Survival." *Contemporary Philosophy* 12, 5:7–8.

Burke, Kenneth. 1966. *Language as Symbolic Action: Essays on Life, Literature, and Method.* Berkeley: University of California Press.

Burnham, Frederic B., ed. 1989. *Postmodern Theology: Christian Faith in a Pluralist World.* San Francisco: HarperSanFrancisco.

Bush, George. 1991. "Making It America's Business to Improve the Environment." *Trilogy* 3, 3:15–18.

Callicott, J. Baird. 1989. *In Defense of the Land Ethic: Essays in Environmental Philosophy.* Albany: State University of New York Press.

Campbell, Joseph. 1988. *Mythologies of the Primitive Hunters and Gatherers.* Part 1 of *The Way of the Animal Powers;* vol. 1 of *Historical Atlas of World Mythology.* New York: Harper & Row.

Caplow, Theodore, Howard M. Bahr, Bruce A. Chadwick, Dwight W. Hoover, et al. 1983. *All Faithful People: Change and Continuity in Middletown's Religion*. Minneapolis: University of Minnesota Press.

Carpenter, James A. 1988. *Nature and Grace*. New York: Crossroads.

Carson, Rachel. 1962. *Silent Spring*. Greenwich, Conn.: Fawcett.

Cheney, Jim. 1989. "Postmodern Environmental Ethics: Ethics as Bioregional Narrative." *Environmental Ethics* 11:117–34.

Clark, William C. 1989. "Managing Planet Earth." *Scientific American* 261, 3:47–54.

Cobb, John B., Jr. 1972. *Is It Too Late? A Theology of Ecology*. Beverly Hills, Calif.: Bruce Publishing.

———. 1979. "Christian Existence in a World of Limits." *Environmental Ethics* 1:149–58.

———. 1990. "The Role of Theology of Nature in the Church." In Birch, Eakin, and McDaniel, eds., *Liberating Life*.

———. 1992. *Sustainability: Economics, Ecology, and Justice*. Maryknoll, N.Y.: Orbis Books.

Cobb, John B., Jr., and Charles Birch. [1981] 1990. *The Liberation of Life: From the Cell to the Community*. Denton, Tex.: Environmental Ethics Books.

Cobb, John B., Jr., and Herman E. Daly. 1989. *For the Common Good: Redirecting the Economy toward Community, the Environment, and a Sustainable Future*. Boston: Beacon Press.

Cobb, John B., Jr., and David Tracy. 1983. *Talking about God: Doing Theology in the Context of Modern Pluralism*. New York: Seabury Press.

Cohen, Michael P. 1984. *The Pathless Way: John Muir and American Wilderness*. Madison: University of Wisconsin Press.

———. 1988. *The History of the Sierra Club, 1892–1970*. San Francisco: Sierra Club Books.

Collingwood, R. G. 1940. *An Essay on Metaphysics*. Oxford: Clarendon Press.

Commoner, Barry. 1971. *The Closing Circle: Nature, Man and Technology*. New York: Alfred A. Knopf.

Conner, John. 1992. "Grassroots Coalition for Environmental and Economic Justice." Unpublished typescript.

Costanza, Robert, ed. 1992. *Ecological Economics: The Science and Management of Sustainability*. New York: Columbia University Press.

Crossman, John Dominic. 1988. *The Dark Interval: Towards a Theology of Story*. Sonoma, Calif.: Polebridge Press.

Cushing, Frank Hamilton. 1979. *Zuñi: Selected Writings of Frank Hamilton Cushing*. Lincoln: University of Nebraska Press.

Dahl, Robert A. 1989. *Democracy and Its Critics*. New Haven and London: Yale University Press.

Daly, Herman E. 1991. *Steady-State Economics*. 2d ed. Washington, D.C.: Island Press.

Daly, Herman E., and John B. Cobb, Jr. 1989. *For the Common Good: Redirecting the Economy toward Community, the Environment, and a Sustainable Future*. Boston: Beacon Press.

Daly, Mary. 1978. *Gyn/Ecology: The Metaethics of Radical Feminism*. Boston: Beacon Press.

———. [1973] 1985. *Beyond God the Father: Toward a Philosophy of Women's Liberation.* Boston: Beacon Press.

Darwin, Charles. [1871] 1952. *The Descent of Man,* in Robert Maynard Hutchins, ed., *Great Books of the Western World,* vol. 49. Chicago: Encyclopedia Britannica.

Davidson, Donald. 1984. *Inquiries into Truth and Interpretation.* New York: Oxford University Press.

Davis, Philip J., and Reuben Hersh. 1986. *Descartes' Dream: The World According to Mathematics.* Boston: Houghton Mifflin.

Dawkins, Richard. 1976. *The Selfish Gene.* New York: Oxford University Press.

———. 1987. *The Blind Watchmaker: Why the Evidence of Evolution Reveals a Universe without Design.* New York: W. W. Norton.

deHaven-Smith, Lance. 1991. *Environmental Concern in Florida and the Nation.* Gainesville: University of Florida Press.

Deloria, Vine, Jr. 1973. *God Is Red.* New York: Dell.

Devall, Bill, and George Sessions. 1985. *Deep Ecology: Living as if Nature Mattered.* Salt Lake City: Gibbs M. Smith.

De Vos, Peter, et al. 1991. *Earthkeeping in the Nineties: Stewardship of Creation.* Rev. ed. Grand Rapids, Mich.: William B. Eerdmans.

DeWitt, Calvin B., ed. 1991. *The Environment and the Christian: What Does the New Testament Say about the Environment?* Grand Rapids, Mich.: Baker Book House.

DeWitt, Calvin B., and Ghillean T. Prance, eds. 1992. *Missionary Earthkeeping.* Macon, Ga.: Mercer University Press.

Diamond, Irene, and Gloria Feman Orenstein, eds. 1990. *Reweaving the World: The Emergence of Ecofeminism.* San Francisco: Sierra Club Books.

Domhoff, G. William. 1967. *Who Rules America?* Englewood Cliffs: Prentice-Hall.

Douglas, David. 1987. *Wilderness Sojourn: Notes in the Desert Silence.* San Francisco: Harper & Row.

Dowd, Michael. 1991. *Earthspirit: A Handbook for Nurturing Ecological Christianity.* Mystic, Conn.: Twenty-third Publications.

Dryzek, John S. 1987. *Rational Ecology: Environment and Political Economy.* New York: Basil Blackwell.

Dunn, Stephen, and Anne Lonergan, eds. 1991. *Befriending the Earth: A Theology of Reconciliation Between Humans and the Earth.* Mystic, Conn.: Twenty-third Publications.

Eco, Umberto. 1990. *The Limits of Interpretation.* Bloomington: Indiana University Press.

Ehrlich, Paul R., A. H. Ehrlich, and J. P. Holdren. 1977. *Ecoscience: Population, Resources, Environment.* San Francisco: W. H. Freeman.

Ehrlich, Paul R., and Anne H. Erhlich. 1990. *The Population Explosion.* New York: Simon and Schuster.

Ehrlich, P. R., and E. O. Wilson. 1991. "Biodiversity Studies: Science and Policy." *Science* 253:758–62.

Einstein, Albert. 1954. *Ideas and Opinions by Albert Einstein*, ed. Carl Seelig, trans. Sonja Bargmann. New York: Crown.

Ekins, Paul, Mayer Hillman, and Robert Hutchison. 1992. *The Gaia Atlas of Green Economics*. New York: Anchor Books.

Eliade, Mircea. 1959. *The Sacred and Profane: The Nature of Religion*, trans. Willard R. Trask. San Diego: Harcourt Brace Jovanovich.

Elmer-Dewitt, Philip. 1992. "Summit to Save the Earth." *Time* (June 1), 42–43.

Engel, J. Ronald, Peter Bakken, and Joan Gibb Engel. 1994. *Ecology, Justice, and Christian Faith: A Guide to the Literature, 1960–1993*. Westport, Conn.: Greenwood.

Erwin, T. L. 1991. "An Evolutionary Basis for Conservation Strategies." *Science* 253:750–52.

Evernden, Neil. 1985. *The Natural Alien: Humankind and Environment*. Toronto: University of Toronto Press.

———. 1992a. "Ecology in Conservation and Conversation." In Oelschlaeger, ed., *After Earth Day*.

———. 1992b. *The Social Creation of Nature*. Baltimore: Johns Hopkins University Press.

Fineman, Howard. 1992. "The Torch Passes." *Newsweek* (November/December Special Election Issue), 4–10.

Firor, John. 1990. *The Changing Atmosphere: A Global Challenge*. New Haven and London: Yale University Press.

Fischler, Martin A., and Oscar Firschein. 1987. *Intelligence: The Eye, the Brain, and the Computer*. Reading, Mass.: Addison-Wesley.

Fleming, Thomas. 1990. "Short Views on Earth Day." *Chronicles: A Magazine of American Culture* 14, 8:7.

Foresta, Ronald A. 1991. *Amazon Conservation in the Age of Development: The Limits of Providence*. Gainesville: University of Florida Press.

Fox, Matthew. 1983. *Original Blessing*. Santa Fe, N.Mex.: Bear.

———. 1988. *The Coming of the Cosmic Christ: The Healing of Mother Earth and the Birth of a Global Renaissance*. San Francisco: Harper & Row.

Fox, Stephen. [1981] 1985. *The American Conservation Movement: John Muir and His Legacy*. Madison: University of Wisconsin Press.

Fraser, J. T., ed. 1981. *The Voices of Time: A Cooperative Survey of Man's Views of Time as Expressed by the Sciences and Humanities*. 2d ed. Amherst: University of Massachusetts Press.

Friedman, Maurice S. 1960. *Martin Buber: The Life of Dialogue*. New York: Harper Torchbook.

Friedman, Milton. 1962. *Capitalism and Freedom*. Chicago: University of Chicago Press.

Frye, Northrop. 1981. *The Great Code: The Bible and Literature*. New York: Harcourt Brace Jovanovich.

Gadamer, Hans-Georg. 1988. *Truth and Method*, ed. and trans. Garrett Barden and John Cumming from 1965 edition. New York: Cross.

Gadon, Elinor. 1989. *The Once and Future Goddess: A Symbol for Our Time*. San Francisco: Harper & Row.

———. 1992. "Metaphors of Birthing: Towards a New Creation for the Age of Ecology." In Oelschlaeger, ed., *After Earth Day*.

Galbraith, John Kenneth. 1952. *American Capitalism: The Concept of Countervailing Power*. Boston: Houghton Mifflin.

———. [1967] 1972. *The New Industrial State*. 2d ed. New York: New American Library.

Geertz, Clifford. 1973. *The Interpretation of Cultures*. New York: Basic Books.

Georgescu-Roegen, Nicholas. 1971. *The Entropy Law and the Economic Process*. Cambridge: Harvard University Press.

Giddings, J. Calvin. 1973. *Chemistry, Man, and Environmental Change: An Integrated Approach*. San Francisco: Harper & Row.

Glacken, Clarence. 1967. *Traces on the Rhodian Shore: Nature and Culture in Western Thought from Ancient Times to the End of the Eighteenth Century*. Berkeley: University of California Press.

Gore, Senator Al. 1992. *Earth in the Balance: Ecology and the Human Spirit*. Boston: Houghton Mifflin.

Gottwald, Norman K. 1979. *The Tribes of Yahweh: A Sociology of the Religion of Liberated Israel, 1250–1050 B.C.E.* Maryknoll, N.Y.: Orbis Books.

Gould, Eric. 1981. *Mythical Intentions in Modern Literature*. Princeton: Princeton University Press.

Granberg-Michaelson, Wesley. 1984. *A Worldly Spirituality*. San Francisco: Harper & Row.

———. 1992. *Redeeming the Creation: The Rio Summit—The Challenge of the Churches*. Geneva, Switzerland: World Council of Churches.

Green, Arthur. 1991. "God, World, Person: A Jewish Theology of Creation, Part I." *Melton Journal*, no. 24.

———. 1992. "God, World, Person: A Jewish Theology of Creation, Part II." *Melton Journal*, no. 25.

Greider, William. 1992. *Who Will Tell the People*. New York: Simon and Schuster.

Griffin, David Ray. 1989. *God and Religion in the Postmodern World: Essays in Postmodern Theology*. Albany: State University of New York Press.

Griffin, David Ray, ed. 1988a. *The Reenchantment of Science: Postmodern Proposals*. Albany: State University of New York Press.

———. 1988b. *Spirituality and Society: Postmodern Visions*. Albany: State University of New York Press.

Griffin, Susan. 1978. *Woman and Nature: The Roaring inside Her*. New York: Harper & Row.

Gup, Ted. 1992. "The Stealth Secretary." *Time* (May 24), 57–59.

Habgood, John. 1990. "A Sacramental Approach to Environmental Issues." In Birch, Eakin, and McDaniel, eds., *Liberating Life*.

Haefele, Edwin T. 1973. *Representative Government and Environmental Management*. Baltimore: Johns Hopkins University Press.

Hardin, Garrett. 1972. *Exploring New Ethics for Survival: The Voyage of Spaceship Beagle*. New York: Viking Press.

Hardin, Garrett, and John Baden, eds. 1977. *Managing the Commons*. San Francisco: W. H. Freeman.

Hargrove, Eugene C., ed. 1986. *Religion and Environmental Crisis*. Athens: University of Georgia Press.

———. 1989. *Foundations of Environmental Ethics*. Englewood Cliffs, N.J.: Prentice Hall.

———. 1992. "Weak Anthropocentric Intrinsic Value." In Oelschlaeger, ed., *After Earth Day*.

Hartshorne, Charles. 1967. *A Natural Theology for Our Time*. LaSalle, Ill.: Open Court.

Haught, John F. 1990. "Religious and Cosmic Hopelessness: Some Environmental Implications." In Birch, Eakin, and McDaniel, eds., *Liberating Life*.

Hays, Samuel P. 1987. *Beauty, Health, and Permanence: Environmental Politics in the United States, 1955–1985*. New York: Cambridge University Press.

———. 1992a. "Environmental Philosophies." *Science* 258:1822–23.

———. 1992b. "Environmental Political Culture and Environmental Political Development: An Analysis of Legislative Voting, 1971–1989." *Environmental History Review* 16:1–22.

Heidegger, Martin. 1962. *Being and Time*, trans. John Macquarrie and Edward Robinson. New York: Harper & Row.

———. 1971. *Poetry, Language, Thought*, trans. Albert Hofstadter. New York: Harper & Row.

———. 1977a. *Martin Heidegger: Basic Writings from* Being and Time (1927) *to* The Task of Thinking (1964), trans. and ed. David Farrell Krell. New York: Harper & Row.

———. 1977b. *The Question Concerning Technology and Other Essays*, trans. William Lovitt. New York: Harper & Row.

Heilbroner, Robert L. 1966. *The Limits of American Capitalism*. New York: Harper & Row.

———. 1972. *The Worldly Philosophers: The Lives, Times, and Ideas of the Great Economic Thinkers*. 4th ed. New York: Simon and Schuster.

Helfand, Jonathan. 1986. "The Earth Is the Lord's: Judaism and Environmental Ethics." In Hargrove, ed., *Religion and Environmental Crisis*.

Hendry, George S. 1980. *Theology of Nature*. Philadelphia: Westminster Press.

Hesse, Mary. 1980. *Revolutions and Reconstructions in the Philosophy of Science*. Bloomington: Indiana University Press.

Hodge, Robert. 1991. *Literature as Discourse: Textual Strategies in English and History*. Baltimore: Johns Hopkins University Press.

Houghton, Richard A., and George M. Woodwell. 1989. "Global Climatic Change." *Scientific American* 260, 4:36–44.

Ihde, Don. 1979. *Technics and Praxis*. Boston: D. Reidel.

Iverson, Peter. 1987. "The Indians of Arizona." In Luey and Stowe, eds., *Arizona at Seventy-five*.

Jakowska, Sophie. 1986. "Roman Catholic Teachings and Environmental Ethics in Latin America." In Hargrove, ed., *Religion and Environmental Crisis.*

Johnston, R. J. 1989. *Environmental Problems: Nature, Economy and State.* London: Belhaven Press.

Jones, Gareth, Alan Robertson, Jean Forbes, and Graham Hollier, eds. 1992. *The Harper-Collins Dictionary of Environmental Science.* New York: HarperCollins.

Kaplan, Abraham. 1963. *American Ethics and Public Policy.* New York: Oxford University Press.

Kaplan, Irving. 1964. *The Conduct of Inquiry: Methodology for Behavioral Science.* San Francisco: Chandler.

Kareiva, Peter M., Joel G. Kingsolver, and Raymond B. Huey, eds. 1992. *Biotic Interactions and Global Change.* Sunderland, Mass.: Sinauer.

Kilday, Anne Marie. 1992. "Bush Backs Change in Wildlife Law." *Dallas Morning News* (September 15).

Kittay, Eva Feder. 1987. *Metaphor: Its Cognitive Force and Linguistic Structure.* New York: Oxford University Press.

Koshland, Daniel E., Jr. 1991. "Preserving Biodiversity." *Science* 253:717.

Krabbe, J. J., and W. J. M. Heijman, eds. 1992. *National Income and Nature: Externalities, Growth and Steady State.* Boston: Kluwer.

Krieger, Murray. 1976. *Theory of Criticism: A Tradition and Its System.* Baltimore: Johns Hopkins University Press.

Kristeva, Julia. [1981] 1989. *Language the Unknown: An Invitation into Linguistics,* trans. Anne M. Menke. New York: Columbia University Press.

Kuhn, Thomas S. 1970. *The Structure of Scientific Revolutions.* 2d ed. Chicago: University of Chicago Press.

Kuo, Cynthia, Craig Lindberg, and David J. Thomason. 1990. "Coherence Established between Atmospheric Carbon Dioxide and Global Temperature." *Nature* 343:709–14.

LaChapelle, Dolores. 1973. *Earth Festivals.* Silverton, Colo.: Finn Hill Arts.

———. 1978. *Earth Wisdom.* Silverton, Colo.: Finn Hill Arts.

———. 1985. "Ritual Is Essential." In Devall and Sessions, *Deep Ecology.*

———. 1988. *Sacred Land, Sacred Sex-Rapture of the Deep: Concerning Deep Ecology and Celebrating Life.* Silverton, Colo.: Finn Hill Arts.

Lakoff, George, and Mark Johnson. 1980. *Metaphors We Live By.* Chicago: University of Chicago Press.

Lappé, Francis Moore. 1990. *Building Citizen Democracy: A Discussion Tool.* San Francisco: Institute for the Arts of Democracy.

Lawson, Hilary. 1985. *Reflexivity: The Post-modern Predicament.* LaSalle, Ill.: Open Court.

Leiss, William. 1972. *The Domination of Nature.* New York: George Braziller.

Lemonick, Michael D. 1991. "War over the Wetlands." *Time* (August 26), 53.

Leopold, Aldo. [1949] 1970. *A Sand County Almanac: With Essays on Conservation from Round River*. San Francisco: Sierra Club Books.

Lerner, Gerda. 1986. *The Creation of Patriarchy*. New York: Oxford University Press.

Levin, Simon A. 1992. "Sustaining Ecological Research." *Bulletin of the Ecological Society of America* 73:213–18.

Lévi-Strauss, Claude. 1983. *The Raw and the Cooked: Introduction to a Science of Mythology*. Chicago: University of Chicago Press.

Lewis, Martin W. 1992. *Green Delusions: An Environmentalist Critique of Radical Environmentalism*. Durham: Duke University Press.

Lincoln, Bruce. 1989. *Discourse and the Construction of Society: Comparative Studies of Myth, Ritual, and Classification*. New York: Oxford University Press.

Lindbeck, George A. 1984. *The Nature of Doctrine: Religion and Theology in a Postliberal Age*. Philadelphia: Westminster Press.

———. 1989. "The Church's Mission to a Postmodern Culture." In Burnham, ed., *Postmodern Theology*.

Linder, Staffan B. 1970. *The Harried Leisure Class*. New York: Columbia University Press.

Livingston, John A. 1981. *The Fallacy of Wildlife Conservation*. Toronto: McClelland and Stewart.

Loftis, Randy Lee. 1992. "Environment Issues Slighted, Critics Say." *Dallas Morning News* (February 1).

Lonergan, Anne, and Caroline Richards, eds. 1987. *Thomas Berry and the New Cosmology*. Mystic, Conn.: Twenty-third Publications.

Lovelock, J. E. 1989. *Gaia: A New Look at Life on Earth*. New York: Oxford University Press.

———. 1990. "Hands Up for the Gaia Hypothesis." *Nature* 344:100–102.

Luckert, Karl W. 1975. *The Navajo Hunter Tradition*. Tucson: University of Arizona Press.

Luey, Beth, and Noel J. Stowe, eds. 1987. *Arizona at Seventy-five: The Next Twenty-five Years*. Tucson: University of Arizona Press.

Lumsden, Charles J., and Edward O. Wilson. 1981. *Genes, Mind, and Culture: The Coevolutionary Process*. Cambridge: Harvard University Press.

———. 1983. *Promethean Fire: Reflections on the Origin of Mind*. Cambridge: Harvard University Press.

Lyotard, Jean-François. 1984. *The Postmodern Condition: A Report on Knowledge*, trans. Geoff Bennington and Brian Massumi. Minneapolis: University of Minnesota Press.

McCann, Michael W. 1986. *Taking Reform Seriously: Perspectives on Public Interest Liberalism*. Ithaca: Cornell University Press.

McCloskey, Donald N. 1985. *The Rhetoric of Economics*. Madison: University of Wisconsin Press.

McConnell, Grant. 1966. *Private Power and American Democracy*. New York: Alfred A. Knopf.

MacCormac, Earl R. 1976. *Metaphor and Myth in Science and Religion*. Durham: Duke University Press.

McDaniel, Jay. 1986. "Christianity and the Need for New Vision." In Hargrove, ed., *Religion and Environmental Crisis.*

———. 1989. *Of God and Pelicans: A Theology of Reverence for Life.* Louisville, Ky.: Westminster/John Knox Press.

———. 1990. *Earth, Sky, Gods, and Mortals: Developing an Ecological Christianity.* Mystic, Conn.: Twenty-third Publications.

———. 1991. "Emerging Options in Ecological Christianity: The New Story, the Biblical Story, and Panentheism." Typescript, Fifth Annual Casassa Conference (March 14–16), Los Angeles.

McIntosh, Robert P. 1985. *The Background of Ecology: Concept and Theory.* New York: Cambridge University Press.

MacIntyre, Alasdair. 1984. *After Virtue: A Study in Moral Theory.* 2d ed. Notre Dame, Ind.: Notre Dame University Press.

McKeon, Richard. 1987. *Rhetoric: Essays in Invention and Discovery*, ed. Mark Backman. Woodbridge, Conn.: Oxbow Press.

McManners, John, ed. 1990. *The Oxford Illustrated History of Christianity.* New York: Oxford University Press.

MacNeil, Jim, Peter Winsemius, and Taizo Yakushiji. 1992. *Beyond Interdependence: The Meshing of the World's Economy and the Earth's Ecology.* New York: Oxford University Press.

Maddox, John. 1992a. "Dangers of Disappointment at Rio." *Nature* 357:265–66.

———. 1992b. "National Academy/Royal Society: Warning on Population Growth." *Nature* 355:759.

———. 1992c. "Unreasonable Hopes for Rio Fades." *Nature* 356:461–62.

Mander, Jerry. 1991. *In the Absence of the Sacred: The Failure of Technology and the Survival of the Indian Nations.* San Francisco: Sierra Club Books.

Marsh, George P. [1864] 1970. *The Earth as Modified by Human Action.* St. Clair Shores, Mich.: Scholarly Press.

Marty, Martin E. 1981. *The Public Church: Mainline-Evangelical-Catholic.* New York: Crossroad.

———. 1989a. "The Establishment That Was." *Christian Century* 106:1045–47.

———. 1989b. *The Westminster Tanner-McMurrin Lectures on the History and Philosophy of Religion at Westminster College, I, 1989: We Might Know What to Do and How to Do It: On the Usefulness of the Religious Past.* Salt Lake City, Utah: Westminster College of Salt Lake City.

Marx, Leo. 1964. *The Machine in the Garden: Technology and the Pastoral Ideal in America.* New York: Oxford University Press.

Massey, Marshall. 1991. "Where Are Our Churches Today? A Report on the Environmental Positions of the Thirty Largest Christian Denominations in the United States." *Firmament* 2, 4:4–15.

Max-Neef, Manfred, et al. 1990. *Human Scale Development: An Option for the Future.* Uppsala: Dag Hammarskjöld Foundation.

Mayr, Ernst. 1982. *The Growth of Biological Thought: Diversity, Evolution, and Inheritance.* Cambridge: Harvard University Press.

———. 1988. *Toward a New Philosophy of Biology: Observations of an Evolutionist.* Cambridge: Harvard University Press.

Meadows, Donella, Dennis Meadows, Jørgen Randers, and William W. Behrens III. 1972. *The Limits to Growth: A Report for the Club of Rome's Project on the Predicament of Mankind.* New York: Universe Books.

Meadows, Donella, Dennis L. Meadows, and Jørgen Randers. 1992. *Beyond the Limits: Confronting Global Collapse, Envisioning a Sustainable Future.* Post Mills, Vt.: Chelsea Green.

Meine, Curt. 1992. "Conservation Biology and Sustainable Societies: A Historical Perspective." In Oelschlaeger, ed., *After Earth Day.*

Merchant, Carolyn. 1980. *The Death of Nature: Woman, Ecology, and the Scientific Revolution.* New York: Harper & Row.

Meyer, Stephen M. 1992. "Environmentalism and Economic Prosperity: Testing the Environmental Impact Hypothesis." Typescript, MIT Project on Environmental Politics and Policy.

———. 1993. "Environmentalism and Economic Prosperity: An Update." Typescript, MIT Project on Environmental Politics and Policy.

———. Forthcoming. *Environmentalism and Economic Prosperity* (provisional title). Cambridge: MIT Press.

Milbrath, Lester W. 1989. *Envisioning a Sustainable Society: Learning Our Way Out.* Albany: State University of New York Press.

Miller, Alan S. 1991. *Gaia Connections: An Introduction to Ecology, Ecoethics, and Economics.* Savage, Md.: Rowman & Littlefield.

Mitchell, Robert Cameron. 1990. "Public Opinion and the Green Lobby: Poised for the 1990s?" In Vig and Kraft. *Environmental Policy in the 1990s.*

Muir, John. 1916. *A Thousand-Mile Walk to the Gulf.* Boston: Houghton Mifflin.

———. 1954. *The Wilderness World of John Muir*, ed. Edwin Way Teale. Boston: Houghton Mifflin.

———. [1938] 1979. *John of the Mountains: The Unpublished Journals of John Muir*, ed. Linnie Marsh Wolfe. Madison: University of Wisconsin Press.

———. [1915] 1979. *Travels in Alaska.* Boston: Houghton Mifflin.

Naess, Arne. [1976] 1989. *Ecology, Community, and Life-Style*, rev. and trans. David Rothenberg. New York: Cambridge University Press.

Nash, James A. 1991. *Loving Nature: Ecological Integrity and Christian Responsibility.* Nashville: Abingdon Press.

Nash, Roderick. 1989. *The Rights of Nature: A History of Environmental Ethics.* Madison: University of Wisconsin Press.

National Commission on the Environment. 1993. *Choosing a Sustainable Future: The Report of the National Commission on the Environment.* Washington, D.C.: Island Press.

Nieswiadomy, Michael L. 1992. "Economics and Resource Conservation." In Oelschlaeger, ed., *After Earth Day.*

Nietzsche, Friedrich. 1872–73. "On Truth and Lies in a Nonmoral Sense." In Bizzell and Herzberg, eds., *Rhetorical Tradition.*

O'Brien, James F. 1988. "Teilhard's View of Nature and Some Implications for Environmental Ethics." *Environmental Ethics* 10:329–46.

Oelschlaeger, Max. 1977. *The Environmental Imperative: A Socio-Economic Perspective.* Washington, D.C.: University Press of America.

———. 1991. *The Idea of Wilderness: From Prehistory to the Age of Ecology.* New Haven and London: Yale University Press.

Oelschlaeger, Max, ed. 1992a. *After Earth Day: Continuing the Conservation Effort.* Denton: University of North Texas Press.

———. 1992b. *The Wilderness Condition: Essays on Environment and Civilization.* San Francisco: Sierra Club Books.

Ogden, Schubert M. 1977. "Prolegomena to a Christian Theology of Nature." In Bemporad, ed., *Rational Faith.*

Olson, Mancur. 1965. *The Logic of Collective Action: Public Goods and the Theory of Groups.* Cambridge: Harvard University Press.

Ortega y Gasset, José. 1932. *The Revolt of the Masses.* New York: W. W. Norton.

Paehlke, Robert C. 1988. *Environmentalism and the Future of Progressive Politics.* New Haven and London: Yale University Press.

———. 1990. "Environmental Values and Democracy: The Challenge of the Next Century." In Vig and Kraft, *Environmental Policy in the 1990s.*

———. 1992. "Environmental Politics and Policy: The Second Wave." In Oelschlaeger, ed., *After Earth Day.*

Paehlke, Robert C., and Douglas Torgerson. 1990. *Managing Leviathan: Environmental Politics and the Administrative State.* London: Belhaven Press.

Pagels, Elaine. 1979. *The Gnostic Gospels.* New York: Vintage Books.

———. 1988. *Adam, Eve, and the Serpent.* New York: Vintage Books.

Passmore, John. 1974. *Man's Responsibility for Nature: Ecological Problems and Western Traditions.* New York: Charles Scribner's Sons.

Paul, Bishop. 1965. *Pastoral Constitution on the Church in the Modern World.* Vatican City, Italy: Libreria Editrice Vaticana.

Peirce, Charles Sanders. 1955. *Philosophical Writings of Peirce,* ed. Justus Buchler. New York: Dover.

Perlin, John. 1989. *A Forest Journey: The Role of Wood in the Development of Civilization.* New York: W. W. Norton.

Pieper, Josef. 1965. *In Tune with the World: A Theory of Festivity.* New York: Harcourt Brace Jovanovich.

Plant, Judith, ed. 1989. *Healing the Wounds: The Promise of Ecofeminism.* Philadelphia: New Society.

Ponting, Clive. 1992. *A Green History of the World: The Environment and the Collapse of Great Civilizations.* New York: St. Martin's Press.

Pope John Paul II. 1987. *Encyclical Letter Sollicitudo Rei Socialis: For the Twentieth Anniversary of Populorum Progressio*. Vatican City: Libreria Editrice Vaticana.

———. 1990. *Peace with God the Creator, Peace with All of Creation*. Vatican City: Libreria Editrice Vaticana.

Pope Paul VI. 1971. *Apostolic Letter of His Holiness Pope Paul VI to Cardinal Maurice Roy: On the Occasion of the Eightieth Anniversary of the Encyclical Rerum Novarum*. (n.p.)

Popper, Karl R., and John D. Ecckles. 1977. *The Self and Its Brain*. New York: Springer International.

Post, Wilfred M., Tsung-Hung Peng, William R. Emanuel, Anthony W. King, Virginia H. Dale, and Donald L. DeAngelis. 1990. "The Global Carbon Cycle." *American Scientist* 78:310–26.

Prance, Ghillean T., ed. 1982. *Biological Diversification in the Tropics: Proceedings of the Fifth International Symposium of the Association for Tropical Biology*. New York: Columbia University Press.

———. 1990a. "The Floristic Composition of the Forests of Central Amazonian Brazil." In Alwyn H. Gentry, ed., *Four Neotropical Rainforests*. New Haven and London: Yale University Press.

———. 1990b. "The Future of the Amazonian Rainforest." *Futures* 22:891–903.

———. 1992. "The Ecological Awareness of the Amazon Indians." In DeWitt and Prance, eds., *Missionary Earthkeeping*. Macon, Ga.: Mercer University Press.

Prigogine, Ilya, and Isabelle Stengers. 1984. *Order Out of Chaos: Man's New Dialogue with Nature*. New York: Bantam Books.

Rappaport, Roy A. 1979. *Ecology, Meaning, and Religion*. Richmond, Calif.: North Atlantic Books.

Ray, Dixy Lee, and Lou Guzzo. 1991. "There Is No Environmental Crisis." In Bernards, ed., *Environmental Crisis*.

Reeves, Richard. 1992. "How Can We Fix Our Collapsing Political Infrastructure?" *Dallas Morning News* (May 18).

Rescher, Nicholas. 1974. "The Environmental Crisis and the Quality of Life." In Blackstone, ed. *Philosophy and Environmental Crisis*.

Ricoeur, Paul. [1975] 1977. *The Rule of Metaphor: Multi-disciplinary Studies of the Creation of Meaning in Language*, trans. Robert Czerny. Toronto: University of Toronto Press.

Roberts, Elizabeth, and Elias Amidon. 1991. *Earth Prayers from around the World: 365 Prayers, Poems, and Invocations for Honoring the Earth*. San Francisco: Harper SanFrancisco.

Rolston, Holmes, III. 1986. *Philosophy Gone Wild: Essays in Environmental Ethics*. Buffalo: Prometheus Press.

Rorty, Richard. 1979. *Philosophy and the Mirror of Nature*. Princeton: Princeton University Press.

———. 1982. *Consequences of Pragmatism (Essays: 1972–1980)*. Minneapolis: University of Minnesota Press.

———. 1989. *Contingency, Irony, and Solidarity*. New York: Cambridge University Press.

———. 1991a. *Essays on Heidegger and Others: Philosophical Papers*, vol. 2. New York: Cambridge University Press.

———. 1991b. *Objectivity, Relativism, and Truth: Philosophical Papers*, vol. 1. New York: Cambridge University Press.

Ruckelshaus, William D. 1989. "Toward a Sustainable World." *Scientific American* 261, 3:166–75.

Rue, Loyal D. 1989. *Amythia: Crisis in the Natural History of Western Culture*. Tuscaloosa: University of Alabama Press.

Ruether, Rosemary Radford. 1975. *New Woman, New Earth: Sexist Ideologies and Human Liberation*. San Francisco: Harper & Row.

———. 1983. *Sexism and God-talk: Toward a Feminist Theology*. Boston: Beacon Press.

———. 1992. *Gaia and God: An Ecofeminist Theology of Earth Healing*. San Francisco: Harper SanFrancisco.

Russell, J. L. 1981. "Time in Christian Thought." In Fraser, ed., *Voices of Time*.

Rust, Eric C. 1971. *Nature: Garden or Desert?* Waco, Tex.: Word Books.

Sagoff, Mark. 1988. *The Economy of the Earth: Philosophy, Law, and the Environment*. New York: Cambridge University Press.

Sahlins, Marshall. 1972. *Stone-age Economics*. New York: Aldine de Gruyter.

Santmire, H. Paul. 1970. *Brother Earth: Nature, God, and Ecology in Time of Crisis*. Camden, N.Y.: Thomas Nelson.

Sayre, Kenneth M. 1991. "An Alternative View of Environmental Ethics." *Environmental Ethics* 13:195–213.

Schaeffer, Francis A. 1968. *The God Who Is There: Speaking Historic Christianity into the Twentieth Century*. Chicago: Inter-varsity Press.

———. 1970. *Pollution and the Death of Man: The Christian View of Ecology*. Wheaton, Ill.: Tyndale House.

Schank, Roger C. 1990. *Tell Me a Story: A New Look at Real and Artificial Memory*. New York: Charles Scribner's Sons.

Schneidau, Herbert N. 1976. *Sacred Discontent: The Bible and Western Tradition*. Baton Rouge: Louisiana State University Press.

Schrödinger, Erwin. 1952. "Are There Quantum Jumps?" *British Journal for the Philosophy of Science* 3:109–10.

———. 1969. *What Is Life? The Physical Aspect of the Living Cell and Mind and Matter*. London: Cambridge University Press.

Schumacher, E. F. 1973. *Small Is Beautiful: Economics As If People Mattered*. San Francisco: Harper & Row.

Schwarz, John C. 1993. "Population, the Church and the Pope." *America* 168, 8:6–10.

Sheldon, Joseph. 1992. *Rediscovery of Creation: A Bibliographical Study of the Church's Response to the Environmental Crisis*. Metuchen, N.Y.: Scarecrow Press.

Shepard, Paul. 1982. *Nature and Madness*. San Francisco: Sierra Club Books.

Simon, Julian L., and Herman Kahn. 1984. *The Resourceful Earth—A Response to Global 2000.* Oxford: Basil Blackwell.

Smith, Adam. [1776] 1952. *The Wealth of Nations.* In Robert Maynard Hutchins, ed., *Great Books of the Western World*, vol. 39. Chicago: Encyclopedia Britannica.

Snyder, Gary. 1974. *Turtle Island.* New York: New Directions.

———. 1990. *The Practice of the Wild.* San Francisco: North Point Press.

———. 1992. *No Nature: New and Selected Poems.* New York: Pantheon.

Soulé, M. E. 1991. "Conservation: Tactics for a Constant Crisis." *Science* 253:744–50.

Spencer, Robert F., Jesse D. Jennings, et al. 1977. *The Native Americans: Ethnology and Backgrounds of the North American Indians.* 2d ed. New York: Harper & Row.

Spitler, E. E. 1992. "The Energy Business and Conservation." In Oelschlaeger, ed., *After Earth Day.*

Splash, Clive L. "Economics, Ethics and Long-Term Environmental Damages." *Environmental Ethics* 15:117–32.

Spretnak, Charlene. 1986. *The Spiritual Dimension of Green Politics.* Santa Fe, N.Mex.: Bear.

Stanley, Manfred. 1978. *The Technological Conscience: Survival and Dignity in an Age of Expertise.* Chicago: University of Chicago Press.

Starhawk. 1979. *The Spiral Dance: A Rebirth of the Ancient Religion of the Great Goddess.* San Francisco: Harper & Row.

———. 1982. *Dreaming the Dark.* Boston: Beacon Press.

———. 1989. "Feminist, Earth-based Spirituality and Ecofeminism." In Plant, ed., *Healing the Wounds.*

———. 1990. "Power, Authority, and Mystery: Ecofeminism and Earth-based Spirituality." In Diamond and Orenstein, eds., *Reweaving the World.*

Steidlmeier, Paul. 1993. "The Morality of Pollution Permits." *Environmental Ethics* 15:133–50.

Stone, Christopher D. 1974. *Should Trees Have Standing? Toward Legal Rights for Natural Objects.* Los Altos, Calif.: William Kaufmann.

———. 1975. *Where the Law Ends: The Social Control of Corporate Behavior.* San Francisco: Harper & Row.

Swimme, Brian. 1988. "The Cosmic Creation Story." In D. Griffin, ed., *Reenchantment of Science.*

Swimme, Brian, and Thomas Berry. 1992. *The Universe Story: From the Primordial Flaring Forth to the Ecozoic Era—A Celebration of the Unfolding of the Cosmos.* San Francisco: HarperCollins.

Tawny, R. H. [1926] 1947. *Religion and the Rise of Capitalism: A Historical Study.* New York: New American Library.

Taylor, John. 1990. "The Future of Christianity." In McManners, ed., *Oxford Illustrated History of Christianity.*

Teilhard de Chardin, Pierre. 1959. *The Phenomenon of Man*, trans. Bernard Wall. New York: Harper & Row.

———. 1965a. *The Divine Milieu*. New York: Harper and Row.

———. 1965b. *Hymn of the Universe*, trans. Simon Bartholomew. New York: Harper & Row.

Thoreau, Henry David. [1854] 1962. *Walden and Other Writings by Henry David Thoreau*. New York: Bantam Books.

Toulmin, Stephen. 1982. *The Return to Cosmology: Postmodern Science and the Theology of Nature*. Berkeley: University of California Press.

Tracy, David. 1987. *Plurality and Ambiguity: Hermeneutics, Religion, Hope*. San Francisco: Harper & Row.

Tracy, David, and John B. Cobb, Jr. 1983. *Talking About God: Doing Theology in the Context of Modern Pluralism*. New York: Seabury Press.

Troeltsch, Ernst. [1912] 1987. *Protestantism and Progress: The Significance of Protestantism for the Rise of the Modern World*. Philadelphia: Fortress Press.

Tucker, William. 1982. *Progress and Privilege: America in the Age of Environmentalism*. Garden City, N.Y.: Doubleday.

Turbayne, Colin Murray. [1962] 1971. *The Myth of Metaphor*. Rev. ed. Columbia: University of South Carolina Press.

Turk, Amos, Jonathan Turk, Janet T. Wittes, and Robert Wittes. 1974. *Environmental Science*. Philadelphia: W. B. Saunders.

United States Bishop's Statement. 1991. "Renewing the Earth: An Invitation to Reflection and Action on the Environment in Light of Catholic Social Teaching." *Origins: CNS Documentary Service* 21:425–32.

Victor, David G. 1991. "How to Slow Global Warming." *Nature* 349:451–56.

Vig, Norman J., and Michael E. Kraft. 1990. *Environmental Policy in the 1990s: Toward a New Agenda*. Washington, D.C.: C. Q. Press.

Weber, Max. [1904–5] 1958. *The Protestant Ethic and the Spirit of Capitalism*, trans. Talcott Parsons. New York: Charles Scribner's Sons.

———. 1973. *Essays in Sociology*, ed. and trans. H. H. Gerth and C. Wright Mills. New York: Oxford University Press.

Weinsheimer, Joel. 1991. *Philosophical Hermeneutics and Literary Theory*. New Haven and London: Yale University Press.

Weizsäcker, Carl Friedrich von. 1988. *The Ambivalence of Progress: Essays on Historical Anthropology*. New York: Paragon House.

Wenz, Peter S. 1988. *Environmental Justice*. Albany: State University of New York Press.

West, Bruce J. 1985. *An Essay on the Importance of Being Nonlinear*. New York: Springer-Verlag.

White, Lynn, Jr. 1967. "The Historical Roots of Our Ecologic Crisis." *Science* 155:1203–7. Reprinted in Barbour, ed., *Western Man and Environmental Ethics*.

———. 1973. "Continuing the Conversation." In Barbour, ed., *Western Man and Environmental Ethics*.

Whitehead, Alfred North. [1929] 1979. *Process and Reality: An Essay in Cosmology*. Corrected edition. Ed. David Ray Griffin and Donald W. Sherburne. New York: Free Press.

Wigley, T. M. L., and S. C. B. Raper. 1992. "Implications for Climate and Sea Level of Revised IPCC Emissions Scenarios." *Nature* 357:293–300.

Wilkinson, Loren, ed. 1980. *Earthkeeping: Christian Stewardship of Natural Resources.* Grand Rapids, Mich.: William B. Eerdmans.

Wills, Garry. 1990. *Under God: Religion and American Politics.* New York: Simon and Schuster.

Wilson, Edward O. 1978. *On Human Nature.* Cambridge: Harvard University Press.

———. [1975] 1980. *Sociobiology: The Abridged Edition.* Cambridge: Harvard University Press. Originally published as *Sociobiology: The New Synthesis.*

———. 1984. *Biophilia.* Cambridge: Harvard University Press.

———. 1992. *The Diversity of Life.* Cambridge: Belknap Press of Harvard University Press.

Wilson, Edward O., and Charles J. Lumsden. 1981. *Genes, Mind, and Culture: The Coevolutionary Process.* Cambridge: Harvard University Press.

———. 1983. *Promethean Fire: Reflections on the Origin of Mind.* Cambridge: Harvard University Press.

Wilson, Edward O., and F. M. Peter, eds. 1988. *Biodiversity.* Washington, D.C.: National Academy Press.

Wittgenstein, Ludwig. 1958. *Philosophical Investigations.* 2d ed. Trans. G. E. M. Anscombe. New York: Macmillan.

———. 1980. *Culture and Value,* trans. Peter Winch. Chicago: University of Chicago Press.

Woodwell, G. M., J. E. Hobbie, R. A. Houghton, J. M. Melillo, B. Moore, B. J. Peterson, and G. R. Shaver. 1983. "Global Deforestation: Contribution to Atmospheric Carbon Dioxide." *Science* 222:1081–86.

World Commission on Environment and Development. 1987. *Our Common Future.* New York: Oxford University Press.

World Council of Churches. 1988. "Liberating Life: A Report to the World Council of Churches." In Birch, Eakin, and McDaniel, eds., *Liberating Life.*

Worster, Donald. [1977] 1985. *Nature's Economy: A History of Ecological Ideas.* Cambridge: Cambridge University Press.

———. 1985. *Rivers of Empire: Water, Aridity, and the Growth of the American West.* New York: Pantheon.

Wright, Will. 1992. *Wild Knowledge: Science, Language, and Social Life in a Fragile Environment.* Minneapolis: University of Minnesota Press.

Yancey, Philip. 1991. "A Voice Crying in the Rain Forest." *Christianity Today* (July 22), 26–28.

Zimmerman, Michael. 1992. "The Future of Ecology." In Oelschlaeger, ed., *After Earth Day.*

———. 1993. "Rethinking the Heidegger–Deep Ecology Relationship." *Environmental Ethics* 15:195–224.

Index